MAN, METALS and MODERN MAGIC

by

J. Gordon Parr

Associate Professor of Metallurgy
University of Alberta
Canada

GREENWOOD PRESS, PUBLISHERS
WESTPORT, CONNECTICUT

Library of Congress Cataloging in Publication Data
Parr, James Gordon, 1927-
Man, metals, and modern magic.

Reprint of the ed. published by American Society for Metals, Cleveland.
Includes index.
1. Metallurpy--History. 2. Metals--History. I. Title.
[TN615.P3 1978] 669'.09 77-25186
ISBN 0-313-20122-6

Reprinted in 1978 by Greenwood Press, Inc.,
51 Riverside Avenue, Westport, Conn. 06880

Printed in the United States of America

10 9 8 7 6 5 4 3 2 1

Acknowledgments

THIS BOOK, however mean its parent, has distinguished — if unwitting — godfathers. For during the past ten years I have taken freely from the knowledge of my colleagues at this and other universities, and gathered opinions about metals and history; and, of course, I have benefited from other authorities whose books and papers I have been able to thumb through. However, any errors and omissions should not be visited upon the godfathers.

The editorial staff of the American Society for Metals has treated me with an undeserved courtesy and, usually, with some patience. Its editorial consultant, Mr. Myron Weiss, has been entirely responsible for collecting the illustrations; and I am grateful to him for doing this job so very much better than I would have done.

I cannot adequately thank my good friend Kenneth Allott of the University of Liverpool, England. Without his encouragement the book would probably not have been written; and even if attempted, it would not have given me such pleasure in writing.

J. GORDON PARR

Table of Contents

Introduction

One of G. K. Chesterton's *Father Brown* detective stories is about a postman who walked into a block of flats, murdered one of the residents, and carried out the corpse in his sack without so much as rousing the suspicion of the commissionaire who stood on duty. "Nobody ever notices postmen, somehow," said Father Brown, and he echoed the paradox that Chesterton repeatedly expounded: the most familiar things can go by most easily unnoticed. Indeed, the paradox itself has now become sufficiently self-evident to be regarded rather as a truism.

We have become so familiar with the modern use of metals that we tend to accept it — like Father Brown's postman — fairly thoughtlessly. Perhaps this is as well: life is short, and there are more invigorating topics to think about. But the present-day use of metals is only the end of a long path of development which cuts across 6000 years of world history, and reveals a section which is certainly not familiar enough to excuse a thoughtless acceptance. The section shows not only the different purposes that metals have served throughout the ages, but, because of their vital consequence to man's life and habits, it discloses the sweeping effects of metallurgical development upon world history.

As metals have become increasingly important, so their study has necessarily become more specialized, and it is difficult for the metallurgist — let alone the layman — to appreciate all that his subject involves or has involved. The metallurgist's microscope symbolizes his enforced practice of examining a tiny part of his subject at terrific magnification. It is time to turn the microscope round (if only its optics will let us) and, as we peep down the wrong end of it, we see today's metals fall into the sobering perspective which 6000 years provide.

CHAPTER 1

Beginning of Metallurgy

It is often pointed out that man existed for a very long time on the earth before he discovered the uses to which he might put metals. Although this is quite indisputable it is hardly remarkable. The rapid advances that we are accustomed to today depend upon the discoveries that were made much more laboriously in the past. And in the first place, because of his lack of experience, man's development was bound to come slowly.

We can take as our starting point the first Ice Age. About 500,000 years ago, for that is when the first glacial period is reckoned to have occurred, there was a creature on the earth that we recognize as being something like a man. He could stand upon his hind legs, and so was able to use his front limbs for other, more interesting, purposes. His physical change resulted in a mental development, for being in a position to use his hands, he was obliged to think; his brain grew.

DIORAMA SHOWING PRIMITIVE MAN BEATING NATIVE COPPER INTO WEAPONS AND IMPLEMENTS.

Courtesy: Copper and Brass Research Association

Direct evidence of these primitive man-like creatures is rare, because firstly, there were probably few of them in widely scattered localities, and secondly, their remains were rarely preserved. Anything edible, including bones, was likely to be eaten up by animals, who, like the younger generation, had bellies but no palates. But primitive man left more abundant evidence of his development in the implements which he made for himself from flints, and which altered in form and improved as the millenia went by.

The idea of making a tool by chipping and shaping a flint must have taken thousands of years to mature, for it involved not only the animal intelligence that prompts a monkey to throw a coconut, but demanded the logical reasoning that allows an end to be achieved by first performing a seemingly unconnected operation. In this case the aim was to kill, and the intermediate action was to make a weapon. This dawn of thinking, crude though it was, marks the beginning of the Stone Age; and the improvements in the way flints were fashioned provide a useful yardstick by which the progress of very primitive man can be computed. This view has not always been held, for until the 17th century at any rate, it was a popular belief that chipped stones were sent from the heavens; and in 1649 Tollius insisted that they were "generated in the sky by a fulgurous exhalation conglobed in a cloud by the circumposed humor." Our vague theories of today are, unhappily, less elegantly expressed.

About 100,000 years ago Piltdown man was once supposed to have existed. But although the scant evidence offered by some remains of a creature found in Sussex has since been written off as a hoax, the flint implements found in the vicinity indicate that there had already been advances made in working stone. The flints are no longer just crudely chipped, but they may be distinguished as tools for a specific use — such as knives, or axes, or borers.

Clearly the ability of being able to think logically (arriving at a solution to a problem) is developing.

There is still one more stage in this history before our direct ancestor comes into the picture — Neanderthal man, who was so similar to man as we know him that he might be regarded as human. But the ethnologists maintain that this is not so: that he is a different species of the same genus; that his skull was not of human pattern, nor was the thumb opposable to the fingers. Nevertheless Neanderthal man made fires — an extremely important advance; he sheltered from the weather in caves, and was right-handed.

Then, after the worst of the last Ice Age had passed, 35,000 years ago, Neanderthal man was displaced from the dominant position he held by another creature — our first ancestor. *Homo sapiens* was similar in many respects to the Neanderthaler. But he could talk, he was more intelligent, and his skull and digits were very like our own.

It is at this point that the real history of mankind begins; and it is here, too, that metals make their appearance in our story, for during the latter part of the Stone Age metals were known and appreciated both for their ornamental and utilitarian value.

Gold, silver and copper are the metals that most commonly occur in the native form; that is, they occur *as* metal, and not in the usual guise of chemical compound. The most important sources of metals (except gold) do not occur in this convenient way, but as complex minerals whose treatment to yield metal may be intricate and expensive.

Of these three native metals, it is likely that gold

was the first to be discovered, for it is the most widely distributed. As it is very resistant to corrosion, the bright surface of a gold nugget would attract the attention of an observant primitive man, and no doubt kindle his disgust when it would not chip or break as other stones did. However, its appearance was pleasing, and the metal was so ductile that, although it was not a first-rate material for armament, it made a pretty ornament and might even be temporarily sharpened into a blade. The Incas, when their American country was invaded by the Spanish in the 15th and 16th centuries, were found to have almost no metallurgical knowledge; yet they used native metals very largely. Incredible as it may seem today, many of their axes and knives were of gold, which they valued much as we value iron — for its utility, not because it is scarce.

But native gold and, more particularly, native silver could not be used a great deal, for their rarity and their softness precluded wide application. Copper, however, was found much more commonly; and although fairly easily corroded, the shining metal surface might be revealed by stream erosion, or by fortuitous inquisitive scrapings. Once discovered, and once some significance had been attached to the discovery, the locality would be searched for more; and whereas the widely distributed metal gold was found as small nuggets or disseminated in quartz rocks, native copper might be discovered as very large masses, probably too big to be moved. A lump of copper hacked off would be beaten and hammered until it became a suitable shape. And during this process it became harder.

This phenomenon, of work hardening, occurs in all metals, but only below a characteristic temperature, which for some metals is somewhat below atmospheric. Fortunately for primitive man, copper hardens at ordinary air temperatures; and after hammering it could be sharpened to make a blade, or pointed to make a spike. By our standards, of course, the metal would still be fairly soft, but it was hard enough for many purposes; and it had the advantage over stone and flint that it was much more easily worked into shape. It is doubtful whether prehistoric copper weapons were hardened by any other process; as far as it is possible to tell, both native copper and smelted copper (which will be described later) were hardened only by this means.

The period during which man used stone and metal by manhandling them into shape in much the same way was a very long one. In the most civilized localities it ended only about 6000 years ago; in some restricted areas it has not ended yet. And 500 years ago there were vast parts of the world where the inhabitants had not been introduced to more advanced techniques. Before the Metal Age began there were two vital discoveries to be made: that metals could be melted and cast to form the impression of a mold; and that metals could be produced from mineral-bearing rocks.

Before we can satisfactorily follow the early history of metals, it will be necessary to define some of the terms that have to be used. While I have no intention of launching into the depths of technicalities and metallurgical jargon — these are reserved for the occasions when an evasive course has to be followed—there are one or two popular misconceptions of the meanings of "mineral," "ore" and so on, that are worth putting right. This might be done in a list of definitions forming an appendix that no one would ever bother to read; or the definitions might be subtly introduced into the text, and too easily overlooked. Alternatively, throwing continuity to the winds, one can digress and say: here are some necessary definitions. And that is the practice that will be followed.

A mineral is any metallic compound that occurs in the earth's crust. This is a fairly broad interpretation, and means virtually that anything in the earth's crust that is not vegetable or animal, as the party game implies, is mineral. If a particular mineral deposit occurs in sufficient abundance to be profitably mined and treated to extract its metallic component, then it is called an ore. The worthless material that somehow or other has to be expelled during the process of extracting the metal from its ore is known as "gangue." For example, ironstone may occur as a siliceous deposit of which, say, half is iron compounds and the remaining half is silicates of calcium, magnesium, aluminum (and many others beside) which make up the gangue.

The metal is usually produced from the ore by smelting. Broadly speaking this involves heating the ore with carbon (as coke) and fluxes to a high temperature, so that the metallic compounds will be reduced to form liquid metal and the fluxes will combine with the gangue to produce a slag. Slag and metal form two distinct liquid layers, and can be separated. Usually the metal is the heavier, lower, layer. The nature of the ore and a suitably arranged smelting process allow only the desired metal to be produced, although, of course, fairly heavy contamination may occur.

It may be economic to re-treat the slag to recover any valuable constituents; and it may also be advisable to refine the metal, producing a more marketable product and perhaps incidentally recovering some valuable metallic impurities. In this way much of the silver, gold and platinum of the world is produced during the refining of copper.

Having cleared away enough of this undergrowth to see the path once again, we can continue with the affairs of primitive man, who, by dint of groping and floundering, eventually found these things out. There is some doubt whether he first discovered that ores could be smelted to produce metals, or that the native metal he had known so long could be melted. It does seem, however, in the light of the metal articles he left behind, that smelting was first known; and this is not surprising in view of the accidental way in which the discovery must almost certainly have been made.

Copper was the first metal to be extracted from its ores. And the most popular of the current theories of primitive man's discovery is that by pure chance he built his hearth of rocks containing a copper mineral which was reduced to metal by the charcoal produced in the fire. Now, it is unlikely that such an occurrence would be noticed immediately; and if it were, there is little chance that much significance would be attached to the phenomenon. But in a copper-rich region it would inevitably happen many times, and eventually some association would be made — it is supposed — between fire and the fused copper produced from rock of a particular kind. In this, by the way, lies the strength of the theory as opposed to rival theories, for it is more than unlikely that a single chance production of metallic copper would have attracted much attention. The pretty story — and it is one of a number — of the Egyptian lady who accidentally dropped her cosmetic, which was made of malachite (copper carbonate), into a charcoal fire and observed its reduction to metallic copper is doubtful on that score alone. What is more, one cannot easily believe that an Egyptian lady so intent on obscuring the work of nature, would be so capable of revealing it. The camp fire theory is more plausible.

The repeated production of copper under these conditions made its mark on the brain of primitive man; metal could be extracted from its ores. Such a happy accident could hardly be responsible for leading him to melt native metal. A man, however primitive, could not be expected to throw his valuable copper weapons, or his wife's equally treasured gold ornaments, on his camp fire, nor be credited with the intelligence of noticing what happened if he did. Most certainly he would not repeat the operation so many times that it would make any impression upon him.

So far, we have avoided dates and locations: the former, because of the recognized difficulty of dating prehistoric events; the latter, because the absence of evidence of metal workings in certain parts of the world is not conclusive of metallurgical inactivity. There may well have come a time when man was so hard-pressed for materials that he melted and recast his earlier weapons and tools; we do it today, without giving a thought to the difficulties we are making for posterity.

GOLD HELMET, MADE BEFORE 3000 B.C., AND RECOVERED FROM THE TOMB OF MES-KALAM-DUG, AT UR.
Courtesy: University of Pennsylvania Museum

The earliest copper workings, which may be dated about 6000 B. C., have been discovered in parts of the Middle East, particularly around Ur in the Fertile Crescent, where it seems likely that the art of metallurgy, like agriculture, was first practiced. Here, probably, at Sumer in Mesopotamia, (the land of Shinar mentioned in *Genesis*), the first civilized community existed, for the silt deposited from the Tigris and the Euphrates as they entered the Persian Gulf produced a fertile land which attracted the peoples of the northern reaches of the rivers. As more immigrants settled it became necessary to organize a scheme of irrigation; and an agricultural community developed, which, in addition to growing crops, could spin and weave and make pottery. It had its wars too; but the weapons were made of stone.

Then came the Great Flood, supposedly about

4000 B. C., destroying all the low-lying farms and villages, exterminating the inhabitants, and laying waste to all that had been built. After the waters had subsided, and the face of the ground was dry, the fertile land again attracted the inhabitants of the north, who, in addition to reviving the agriculture of the delta, introduced new arts. Among these were the arts of melting and casting copper, silver and gold, and of making copper from its ores.

The ornaments and weapons found during the excavations at Ur imply that the casting of metal had begun about 3500 B. C. And it seems that in this locality there was an interval of about 2000 years between the first crudely hammered metal article and the earliest cast one.

Remains of primitive copper workings have been found at the sites of other early civilizations — in the Nile valley in Egypt; and at Mohenjo-daro, which appears to have been the center of an early Indian culture at Sind. A complete survey of this subject would fill several books. But there seems to be no evidence to suggest that in any of these places metallurgy was more advanced than in Sumer. The general progress of the art of producing and using metals followed much the same sequence. It is only the speed of the advance and the time over which each stage of development lasted that differed very greatly.

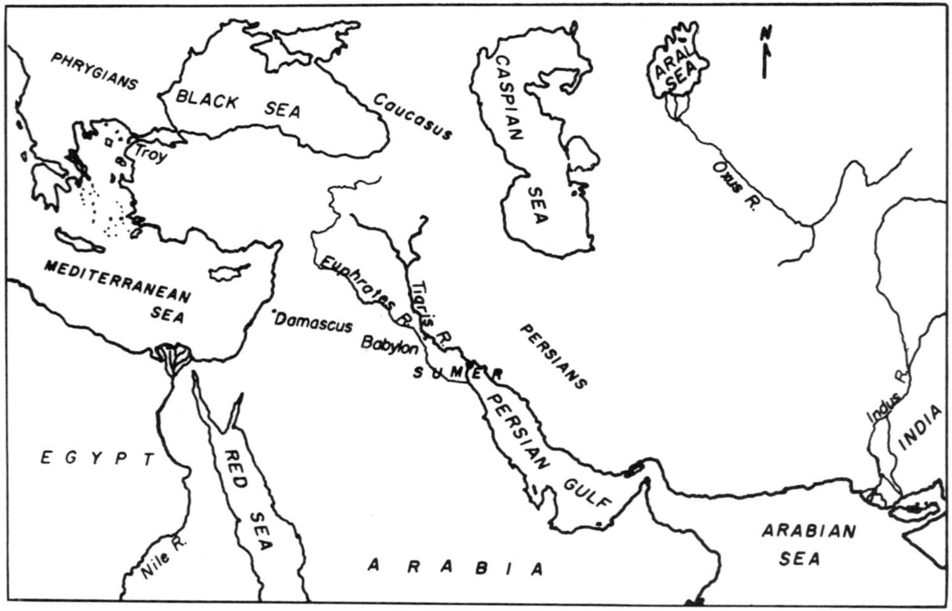

MAP OF ANCIENT WORLD.

CHAPTER 2

4000–1400 B.C.

THE CHINESE, who are credited with so many proverbs, are said to have the adage that "the shortest step may be the beginning of the longest journey." It was the smallest conceivable advance that led primitive man to make a shallow cavity in the hearth of his fire to collect the metal that he had learned to make; and a long eventful journey through the ages that has reached the day when furnaces can smelt 2000 tons of charge each 24 hours; a long eventful journey, and a fascinating one, so full of incident that few historians have acted as guides through its entirety.

It may seem presumptuous that we should rush in where historians have feared to tread. But following a metal path we should be able to resist the temptations of turning aside and getting lost in the bush. Our object is less to trace world history in its relation to metallurgy, than to outline a history of metallurgy indicating where it has influenced and been influenced by contemporary events.

Metallurgical progress is outlined by the early manufacture of bronzes in the East; the vast (and unfortunate) impetus that the Roman occupation had upon mining; alchemy; the significance of iron making in the Middle Ages; the revival of industry in the 16th and 17th centuries; the metal production of America leading to a world supremacy; steel making and the machine age; and the development of metallurgical science.

The relationship between metallurgical development and our civilization is, of course, an inevitable one, but it is not always possible to differentiate between cause and effect. Clearly, in the very beginning, the discovery of copper led to the use of the metal for implements, and it was not a demand for copper that inspired its discovery. At the other extreme, it was the need for specific engineering materials that often led to the search and the invention of suitable alloys. These were not previously known, and then only later applied. In early times the discovery usually suggested a use, much as a child who finds an empty box will think of something to do with it. In more recent times the demand is prior. Once it is made, sooner or later it will be met — much as a child will demand a model railroad, and, if he makes himself a big enough nuisance, will probably get it. After one demand has been satisfied, more applications may well be found, modifications made, and so on. The child who found the empty box may kick it around the house, or make something of it; the one who has been given a model railroad may stamp on it, or alternatively demand a bigger one. So it is with practical discoveries. They may be used or abused; left idle or developed. Progress has been made by the alternate surging of invention and necessity; a game of scientific cat and utilitarian mouse.

At an early stage in this game, man made a shallow trough at the bottom of his camp-fire hearth, collected the metal there, and then beat it into shape. Gradually he developed this trough into a primitive furnace by digging a small hole, about a foot in diameter, and purposefully smelting metal in it by charging layers of charcoal and copper ore on a charcoal fire burning at the bottom. By making the hearth on the side of a hill and extending it upwards with a stone wall around it, wind encouraged the fire, and the charge melted more rapidly. When all the ore was thought to be reduced, any slag and unburnt charcoal were raked off and the metal cooled. As soon as it had become solid, the copper cake was pulled out and broken up while it was still hot and brittle, and the pieces of copper were hammered to make suitable implements. Analyses of such prehistoric copper show a complete absence of sulphur, indicating the use of oxidized ores or native metal. This is what one would expect, since the oxidized ores of copper occur at, or near, the surface and are the most readily available.

These advances led only to a more thorough exploitation of the natural resources of metal; and smelted copper was used in much the same way as native metal had been. But it was not long, presumably, before the idea of casting was born. It became apparent that the lower surface of the metal cake assumed the shape of the hearth; and by making the hole with suitable contours, it became possible to reduce the amount of work that was needed to chip and hammer the metal into shape. Simple forms like axes and blades could be cast; but they were limited in value, as their upper surfaces were bound to be flat. Then castings were made in simple impressions cut into rock or pressed into clay, and before long, hollow molds were made of two pieces of stone held together so that a casting could be made that was shaped on all sides.

Simultaneously, improvements were made in the smelting process. The hearth became bigger and relied more upon wind to stimulate the process that went on within it; and certainly before 1500 B. C. bellows were introduced to produce a direct blast. The crude copper was refined by melting it again in a deep hearth lined with clay, when the clinker and dirt rose to the surface of the liquid metal and were skimmed off, leaving a cleaner, purer material. Eventually clay crucibles were made which dropped into the hearth and which could be lifted out so that metal might be poured into molds. Before the idea of a crucible was known it is doubtful just how molten metal was transferred from hearth to mold. Presumably, a hot stone ladle or a suitably cut channel was used.

In spite of this progress, stone implements were still widely used, for copper is a soft metal. Prehistoric man, however, needed a harder, stronger metal; and in some localities he was lucky enough to produce one owing to the nature of the ores he used. In these favored places the ores contained minerals in addition to those of copper, which were simultaneously reduced during the smelting process; and notable among these was cassiterite (tin oxide). Consequently, by accident rather than intention, an alloy of copper and tin was produced, rich in copper and something like the material we now know as bronze. Its advantages were appreci-

EGYPTIAN COPPER EWERS, MADE ABOUT 2800 B.C.
Courtesy: Copper and Brass Research Association

ated: it was harder and would more readily work harden than copper; and it was a better metal to cast, producing a sounder casting with a more faithful impression. Before long it was realized that the alloy could be made by smelting together certain ores from different localities – copper ores and tin ores. But for a long while why the "copper" that was made in this way was so superior remained a mystery; and it was only quite recently that man learned to make bronze by melting metallic tin and copper together. As the proportion of tin in a bronze is quite critical (to give adequate properties for various purposes there should be between 1 and 10% of tin) the prehistoric production of bronze must have been a chancy business. While we can get some idea of the extent of bronze culture by the number of tools, ornaments and weapons that have been left behind on various sites, there is no indication of the vast number that must have been made, discarded and remelted. The same mystery surrounds many contemporary foundries.

Tin ores, however, are not nearly so widely distributed as copper ores. Therefore, in some places the Copper Age was a prolonged one, and stone implements were used throughout most of it.

The bronze culture of Egypt was comparatively slow in developing, for the Egyptians' source of copper was the Sinai Peninsula, where tin did not occur. On the other hand, the use of bronze in Sumer began much earlier. In some areas the Bronze Age began so late as to be almost non-existent, for as a result of the knowledge brought by invaders or immigrants, the use of iron soon followed that of copper. This was so in Peru, where bronze was only just beginning to be made in the 16th century when the Spaniards arrived; and in North America iron follows copper without any Bronze Age supervening.

The opening of the Bronze Age in the Ancient World not only represented the first production of a truly useful metal, but, because of the comparative scarcity of sources of tin, led to trading. The inhabitants of the second city of Hissarlik (the site of Troy) learned the art of bronze making about 2000 B.C., and discovered, too, that copper and tin could be obtained from Central Europe. Egypt, without tin and therefore without bronze, traded with Hissarlik, and also with Crete. Further, about this time, bronze implements — probably from Hissarlik — reached Portugal and Spain; and soon afterwards knowledge of the alloy spread along the coast of Western Europe to Great Britain.

About 1900 B. C. Hissarlik II was burned down, and its commerce necessarily ended. Consequently, Cretan trade assumed greater importance. Much of this trade was with Egypt, which, however, about 1800 B. C. suffered a series of political misfortunes and eventually dropped out of the picture. As an outlet for their merchandise the Cretans were forced to find markets in Sicily, Spain and southern Italy — a process which ended with the exhaustion of their available tin. This was a fate that in turn was to come to many countries.

But by this time the Middle Europeans and the Spaniards had learned how to make bronze, and the cessation of supplies from the East acted as a stimulus to native industry. Because of Spanish manufacture, more bronze implements found their way into Britain, via the coast of France. But very soon the Spanish ores temporarily came to an end, for in those days only the most accessible deposits could be used, and hidden ore bodies were neither discovered nor exploited. Of necessity the Bronze Age began in Britain.

Another dissemination of bronze culture was brought about by the drought, about 1600 B. C., that compelled the inhabitants of Turkestan (that country to the east of the Caspian Sea) to migrate to Mesopotamia and Egypt, and also to parts of Europe, India and possibly China. They had learned

to make and use metals; and their domestication of the horse gave them a superiority which enabled them to form powerful settlements or dominate communities whose ways were more primitive.

The purpose of this brief account is to show how the advent of the Bronze Age increased the tempo of man's life. Before it, trading was almost unknown; exploration was inspired only by the fertility of the land, and by the animal instincts that prompted battle. Then, in a comparatively short space of time, the world was no longer so obviously divided into East and West, and by about 1800 B.C. there were centers of bronze-making dispersed over much of the known world.

The beginning of the Bronze Age in China is less clearly defined. Most certainly it does not date so far back as the Sumerian Bronze Age, for the earliest Chinese civilizations were in the valley of the Yellow River, where there were no obvious mineral deposits, and the community was essentially an agricultural one. The oldest Chinese bronze objects are supposed to date from 2200 B.C.; and about that time, metal mining began. If one accepts a traditional Chinese belief that the knowledge of bronze came to China from the West, this date is not an improbable one, for in Sumer bronze was known as early as 3000 B.C. The fact that there is little or no evidence of trading may be accounted for by the early exhaustion of raw materials in Sumer. But

SUMERIAN COPPER STATUE OF A MAN CARRYING BRICKS, ABOUT 2600 B.C.

Courtesy: Metropolitan Museum of Art

the art of bronze making may have spread to China from this source; and the Chinese may well have been taught to find and use their own mineral deposits. One must, presumably, believe either that this is possible, or that Chinese metallurgy does not date back so far as 2200 B. C.

Although the Egyptians made a late start in the manufacture of bronze, their technique of casting soon became well advanced. Further, when they first dominated Syria soon after 1400 B. C. sources of tin were made available, and castings in bronze became much more common. No attempt will be made here to give even the barest outline of subsequent events in Syria, Egypt and Sumer (which by this time was part of the Babylonian Empire), for these are as involved when examined in any detail as 20th century European history will appear in a few millenia. H. G. Wells, in his *Short History of the World,* sums them up easily, recording in a comprehensive phrase "the comings and goings of the armies of the Egyptians and the various Semitic powers of Asia Minor, Syria and Mesopotamia." So long as we bear in mind that by 1400 B. C. bronze was made throughout most of the Ancient World, as well as in parts of Europe and China, we need not tax ourselves with historical detail.

In all probability the Egyptians were responsible for the introduction of the *cire perdu,* or "lost wax," method of casting, which allowed metal articles to be made in exact replica of a model. Essentially the process involved then, as it does now, making a solid model in wax to be covered with clay. The clay is hardened by heating it and the wax is melted out, leaving a clay mold the exact negative of the model. Molten metal is then poured in, and allowed to solidify before the mold is broken away to allow the bronze casting to be removed. If a hollow casting was needed, then a core was fashioned roughly the same shape as the object to be made, and was coated with wax which was molded exactly to shape. The clay mold was made over this, the core fastened to it by thin supports, and the wax melted out. Then metal was poured into the space which the wax occupied to produce a hollow replica. It is vital that the core should be firmly held, and the way in which the Egyptians did this is not known with any certainty. In later years iron struts were used; but this method is not very probable before about 1400 B. C. Bronze or copper struts would probably be melted by the metal poured around them; and it is not very easy to imagine stone being used. However, this argument involves a bone of contention that will be gnawed later.

One of the earliest examples of "lost wax" casting is the statue of Pepi I and his son, which is dated about 2600 B. C. It is made of hollow copper or bronze, and the reproduction of detail illustrates how the Egyptians were masters of the art of casting at this early stage.

The danger that metal might solidify before it had filled the mold was appreciated by the Egyptians who prevented this by allowing the metal to run by more than one entry, and also by heating the mold. These precautions are still known to be very necessary today. Intricate castings which could not be made in one piece were fashioned in sections that were joined together by a mortise joint or by a bronze pin; and these devices, too, show a remarkable facility.

The man that could make metals was, as one would expect, a powerful and important figure. His power sometimes led to his being worshipped; alternatively, he might have been hated. In some tribes, apparently, to call a man "a smith" was an extreme form of abusive language. Not surprisingly the metal-makers' craft was associated with magic and wizardry. Perhaps that is how the name "blacksmith" came about. After all, if a man could turn a piece of hard, dull rock into soft, bright metal he was certainly one to be held in respect, or fear. Consequently, the smith's hammer was a sign of power, and an oath taken over his anvil could never be broken. (The importance of a promise made over a blacksmith's anvil was, until recently, well appreciated — or should have been — by eloping couples who fled to the Scottish border where they were married by the smith at Gretna Green.) Further, the smith was supposed to be able to foretell the future by looking at the molten slags in his furnace. Of course, the contemporary metallurgist should be able to foretell the *immediate* future by looking at his slags; he may even foretell the value of some industrial stocks; but the old smith's abilities went further than that.

It is, perhaps, a pity that 20th century metallurgy has lost its supernatural associations. The most we can do in this sort of respect is to fool our admirers with thermodynamic jiggery-pokery. And then we are all too often found out.

GILDED STATUETTE, PROBABLY OF NORTH SYRIAN ORIGIN. THE FIGURE IS 12 INCHES HIGH, AND DATES ABOUT 1500 B.C.

Courtesy: The Oriental Institute of the University of Chicago

The uses to which copper and bronze articles were put during the period 3000 to 1400 B. C. should be considered cautiously. Today, we use aluminum for aircraft; but in 100 years' time we may not. It is not very likely that aluminum planes will be left to slowly corrode away on garbage dumps, and consequently the archeologist of 6000 A. D. may have no clue of what was once a valuable usage of aluminum. He may, however, discover an aluminum hot-water bottle discarded by English settlers and buried in an ancient rubbish dump; and he may find some relics of an aluminum bottle opener, once given away as part of an advertising campaign. If he were hasty in drawing conclusions, he might say that the people of 1957 A. D. used aluminum mainly for hot-water bottles and bottle openers.

It appears that the communities in the Ancient World which practiced the art of metallurgy used bronze and copper for ornaments and for weapons, but this does not preclude the possibility of more extensive uses. The greatest sources of bronze relics of this period lay in the tombs and temples of the rulers of the East, where ornaments and weapons were placed at the side of the dead to help them on their journey. Other remains have, of course, been found. But they are rarer, and so we may get a distorted impression of the variety of purposes to which bronze was put. Yet, owing to limited resources and primitive methods, it is quite possible that bronze was not much more widely used until a later date, when it served for coins and a variety of household pots and pans, as well as for weapons.

Although until 1400 B. C. bronze and copper were

used for the purposes that man's imagination and craftsmanship allowed, other metals were known too. They have been given no place in this account because bronze was by far the most important, both for its utility and because it opened up world trading. The age is properly called a Bronze Age.

However, we have already pointed out that gold

DIORAMA SHOWING EGYPTIANS CASTING THE BRONZE DOORS FOR THE TEMPLE OF KARNAK AT THEBES, ABOUT 1400 B.C.

Courtesy: Copper and Brass Research Association

and silver were known before the discovery of copper and, with the advent of melting and casting practices, these metals were used more extensively for ornamental purposes.

In nature, silver often occurs with lead, whose sulphide mineral (galena) is bright and lustrous. Possibly, then, mineral lead was noticed as early as silver. But evidence seems to indicate that the metal was not made use of until a later period. Shapeless

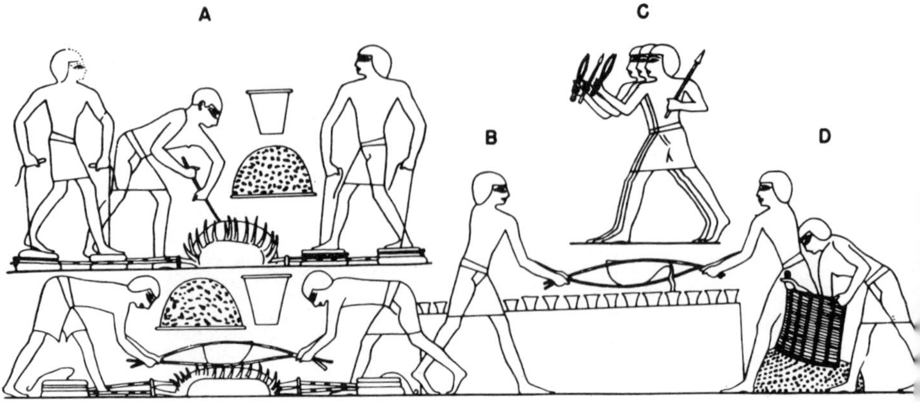

lumps of lead have been found at the site of Hissarlik, at a level dated about 3000 B. C., while an equally old lead relic in the form of a figure was found in the temple of Osiris at Abydos, on the banks of the Nile. The metal, even though it may have been known in restricted localities, was certainly not appreciated, for there is no evidence of a spread of "lead-culture." One cannot be certain whether the metal was smelted or found native, for although lead does not normally occur in its metallic state it is conceivable that it might be produced by an accidental fire and discovered later. If, however, the lead objects of this period were smelted purposefully by man, it is difficult to explain why there are so few of them, and why the value of lead was not exploited. Exploitation came much later in fact, for, although in 1400–1200 B. C. it was possibly used for coins of some sort, it was not until

E

WALL PAINTINGS (ABOUT 1500 B.C.) OF EGYPTIAN METAL-WORKING. *A* Fires are blown to a glow by wind from bellows worked with the feet and with cords held in the hands. In the lower register metal is melted in a crucible held over the fire between two rods. *B* Molten metal poured from crucible into molds. *C* Workmen carrying tongs and blow-pipes. *D* Laborer emptying basket of fuel, probably charcoal. *E* Vessels are beaten into shape by stones on anvils, polished smooth by small pebbles, and ornamented with incised decoration. The fire in the small furnace was probably used for soldering. Courtesy: Metropolitan Museum of Art

about 500 B. C. that its uses became at all extensive.

Our arbitrarily chosen limit of 1400 B. C., up to which date events of metallurgical interests have been outlined, has a significance real, or false, depending upon which side of the fence one stands; for this side of the fence thinks that iron was first produced at this time. No one could disagree that the introduction of the subject of iron deserves a new chapter. Many, however, will dispute the choice of 1400 B. C. as its earliest date of manufacture. An accurate estimate seems to have defeated the archeologists; but for us, 1400 B. C. is as good as any. And to be strictly fair, the cases for and against that selection of date will be put with something approaching impartiality.

CHAPTER 3

1400 – 500 B.C.

By 1400 B. C. the arts of metallurgy, pottery and agriculture were well advanced compared with scientific knowledge. It was generally believed, for example, that the earth was rectangular and the sky held up by mountain peaks; that the stars hung by ropes from the sky and the sun floated round the earth on a never ending river cruise. And, widening the discrepancy between the lack of understanding of the natural sciences and the advanced state of metal culture, iron began to assume an ever growing importance — iron, "the most deadly fruit of human ingenuity." But even before 1400 B. C. its use had not been entirely precluded by any lack of "human ingenuity," because some remains of iron date as far back as 4000 B. C. For, although iron may not have been smelted until much later, and although it does not occur native like gold and copper do, primitive man was provided with small amounts of iron from other than terrestrial sources.

Despite several recorded falls of celestial iron

— meteorites — throughout the ages, many scientists until the 19th century were rather embarrassed and ashamed to have to admit the possibility. Then, in 1803, there was an obligingly spectacular downfall of meteorites in Normandy which was sufficiently uncompromising to convince most of the doubters. So, although there is no truly native iron produced on earth, pieces of the metal have been hurled at us; and some of them must have been discovered by primitive man.

The presumption that primitive peoples used meteoric iron as a rare and valuable material is substantiated by an analysis of their metals. While archeologists, understandably enough, do not encourage metallurgical analysis of their finds, early irons that have been investigated show a high content of nickel, which is peculiar to meteoric irons, and is certainly not a characteristic of iron made directly from ores found in the earth's crust. This high nickel content goes a long way, too, to explain the good state of preservation of most of these old articles — an effect of the presence of nickel, which was profitably rediscovered thousands of years later. Such iron could be hammered and worked just like native gold and copper, and it became harder than either of these. But it was much more highly prized, largely on account of its rarity and because it was more obviously "heaven-sent" than other metals.

In support of the idea that early iron was meteoric in origin it may be pointed out that religious and other ancient writings frequently associate iron with the heavens. Further, later-developing civilizations, when they were first discovered by explorers from more advanced lands, were reported to know and use meteoric iron. The inhabitants of South America, for example, had not learned how to smelt iron when they were subjugated by the Spanish, but they made some implements from the iron of meteorites. And the Eskimos relied upon meteoric iron until as late as the 18th century, when European trade provided them with the smelted material.

On the other hand, it is argued that as iron ores occur very widely, and since the metal is easy to produce from them, it is likely that iron smelting began as soon as copper smelting, or perhaps before. But the smelting of iron is really not so simple. Or at any rate, although it may seem to be today, it would most certainly not have been to primitive man, for a temperature far higher than that achieved in his crude hearth is necessary to make iron which looks anything like a metal. When the simple smelting process as it has been described is applied to iron, it yields only a spongy material with no metallic properties nor any obvious use. It is quite possible that, like copper, iron was produced accidentally many times in the camp fire; but, as it completely lacks luster, it may have gone unnoticed repeatedly. To transform it into a useful metal it is necessary to forge it. That is, the spongy lump has to be hammered and reheated alternately until most of the slag associated with it has been squeezed out. This produces wrought iron, a material which until the introduction of cheap steel in the 1870's A. D. was of first-rate importance, and today still has valuable, but limited, applications. But it is difficult to imagine so elaborate a process being conceived or practiced as early as 4000–3000 B. C.; and irons which pre-date 1400 B. C. do not exhibit the structure that is characteristic of wrought iron. On the contrary, their analysis implies a meteoric origin. However, as comparatively few prehistoric irons of this antiquity have been examined, the second argument has less force than it might have.

Another argument that is put forward in support of early iron smelting is that much of the stone-masonry of pre-1400 B. C. could hardly have been achieved with copper or even bronze tools. If one answers the exponents of this point of view by suggesting that primitive man was more patient than we are, and that, although the process would be tedious, it would not be impossible to carve stone with bronze chisels, they reply most defiantly: "Try it." It is doubtful, too, whether meteoric iron occurred in such profusion as to allow its use for such a commonplace purpose as carving stone. As far as I know, the only answer (an unsatisfactory and dangerous one) is that these early people had very much more patience, and perhaps a little more ingenuity than we credit them with. (But, of course, if we allow them enough ingenuity to carve stone in a way that we cannot imagine now, 6000 years afterwards, perhaps we should credit them with sufficient inventiveness to smelt iron.)

The fact that the few remaining iron implements antedating 1400 B. C. are of a composition that implies meteoric origin does not allow any conclusions to be drawn on this subject, for ordinary man-made wrought iron would, except under the most artificial conditions – even more preservative than the static dry atmospheres of Egyptian tombs — rust away in a comparatively short time.

However, even if one insists that iron was made as long ago as 4000 B. C., it is not until about 1400 B. C. that its use begins to assume importance, and for that reason alone, it is quite justifiable to suggest that this date marks the beginning of the Iron Age.

Probably the earliest known man-made iron article is a dagger blade found in the tomb of Tut-ankh-amen; and this was placed in such a position as implies that it was a most treasured possession. The dagger dates from 1350 B. C., while only 50 years afterwards there is the first known recorded reference to man-made iron. The reference is in a letter written a little after 1300 B. C. by the King of the Hittites, and is of a "sorry-for-the-delay" type, of which he may have been the originator. He wrote to the Egyptian Pharoah, Rameses II, that "there is no good iron in the house of my seal at Kissuwadna, for it is a bad time to make iron. I have written ordering them to make good iron, but so far they have not finished it. When they do, I will send it to thee. Behold! Now I am sending thee an iron dagger blade!"

Digressing for a moment — Jeromewise, putting the pill in the sugar — events of this kind make one aware of the consistency of human nature. The King of the Hittites is slow to give away a material that he could, and did, use to his own advantage. Rameses, whose people expended so much time and effort in honoring him, is the prototype of the modern dictator. He erected statues in his own

CHINESE WINE VESSEL IN BRONZE, 1550–1050 B.C. HEIGHT ABOUT 12 INCHES.

Courtesy: Metropolitan Museum of Art

honor, and inserted his name on many that had been built to commemorate his predecessors. And it was not in this alone that he shone as a builder, for he had more than 150 children, and one does not ask how many wives. In this respect, perhaps, human habits, if not human nature, have changed.

The letter from the King of the Hittites is said to indicate that although iron was probably in very short supply, and that the art was confined to limited areas, differences in quality of the material were already appreciated. It seems more probable, however, in view of the unsettled relations between the Hittites and the Egyptians, that the reference to "a bad time to make iron" was simply a politic excuse. Only later civilizations have been sufficiently sporting to provide their enemies with arms.

To those of us who have no more than a general acquaintance with the affairs of the Ancient World, the origins of ironmaking seem to be wrapped in an almost impenetrable obscurity. The Hittites, between 1500 B.C. and 1400 B.C., were becoming increasingly powerful, due largely to their early knowledge of iron; but whether they discovered or only learned the art of ironmaking at second hand is questionable. It has been supposed for a long time that the art originated somewhere in the mountainous country to the south of the Black Sea, and this is substantiated by the fact that suitable iron ores are found there. A study of the nature of the remains of early iron objects has indicated that appropriate ores were available in the foothills of the mountains to the southeast of the Black Sea, which (together with a lot of Asia Minor) were then dominated by the Hittites. Further, this belief is supported by legend, and together they indicate that iron was first made in the valley of the River Halys, which flows from the mountains to the Black Sea.

The Hittites traded their metal to the Palestinians, for in Gerar implements have been discovered which are believed to date from about 1350 B. C.; and within a hundred years or so, the art of iron-making had been either learned or discovered in Palestine. Furnaces dating to 1200 B. C. have been unearthed at Gerar, together with iron objects that were made there.

So, to avoid the confusion or boredom of detail, we can summarize by asserting that iron was first made in the north of the Ancient World — a belief that is not without its romance, for copper smelting too was introduced into Mesopotamia by settlers from the north. No one knows where these settlers came from; no one is positively certain where iron was first smelted; nor can anyone contradict us if we believe that both originated somewhere in that mountainous country between the Mediterranean and the Caspian Seas.

The way in which these old metallurgists made their iron can only be surmised from the type of metal they produced, and from the methods used by later primitive peoples whose techniques are more adequately recorded in history. In all probability they had discovered the value of building a deeper hearth (either by digging a bigger hole, or by making a cylindrical clay wall around an existing hearth) and so produced a higher temperature and more metal. Charcoal and ironstone were charged on a lighted fire, and a draft was induced by bellows which blew air through clay tubes inserted near the bottom of the charge. After many hours had elapsed, and possibly more charcoal had been added, the furnace was cooled sufficiently for the spongy metal to be removed. The iron was dirty, unmetallic in its appearance; and it was contaminated with clinker and gangue. Therefore, only the cleaner portions were selected, and forged by alternately heating and hammering them. One can only presume that, in the very earliest days of ironmaking, the hot metal would be handled with green wood tongs or stones, and hammered with a heavy stone. In doing this the adherent impurities would detach themselves, and the pockets of earthy slag which were inevitably contained within the metal would be strung out. This improved the mechanical properties enormously, for while a lump of slag produces a region of weakness, and consequently results in easy fracture, strings of slag do not cause such a local rift in the metallic structure. Further, the forging operation reduced the sponginess of the metal, and after sufficient working, the iron could be hammered, still hot, into suitable shapes.

The operation of forging, then, necessitated heating and reheating the iron in a charcoal fire – a process that inevitably resulted in the outside of the metal absorbing carbon. When one is first confronted with the fact that a hot, but solid, element may take into its structure a second element, the concept is difficult to accept. But solid diffusion of this kind does indeed occur. And, in fact, it plays a crucial part in all aspects of metallurgy: in the heat treatment of our modern steels, when carbon atoms diffuse to produce one of many characteristic structures; in iron production, when oxide ore is reduced in the blast-furnace to metal before the charge becomes molten. Then, in the production of the duralumin types of alloys (those metals presently so valuable in aircraft) it is a diffusion of atoms of a number of elements within the metal that produces the effect of "age-hardening" and, consequently, the vital properties of the alloy. The classic experiment which illustrated solid diffusion involved clamping together a flat surface of gold and of copper; and after many years had elapsed, analysis showed that the copper had become contaminated with gold, and *vice versa.* Heating a metal weakens the bonds which hold its atoms together, and so diffusion is quicker at higher temperatures. Yet, although the effects of diffusion are well known, its mechanism is not always clear, and a complete understanding would probably solve many problems that confront the metallurgist today.

But, to return to 1200 B. C., or thereabouts, it was inevitable that carbon should be taken into the outside of the forged iron. Now steel is essentially an alloy of iron with up to about 1.5% of carbon; and, as this is the kind of material that was produced at the surface of forged iron, it is often suggested that as early as 1200 B. C. steel was being made. This is a misleading and almost improper suggestion, for steelmaking involves the production of a uniform, homogeneous material, which this certainly was not. The old ironmaker was, in effect, inadvertently case-hardening his iron in a way that is very similar to modern case-hardening processes for steel. But to suggest that he was *making* steel is as incorrect as to assert that the baby thrashing the piano next door is making music. What is more, he did not at first realize that it was heat rather than the hammer causing the hardening; and it was more than

3000 years later that the part carbon played in the operation was appreciated.

But whether the process was understood is of minor importance. No doubt the man who first discovered the hardening effect thought that he understood it. When it was partly attributed to fire, again it was assumed to be fully understood; and today, when we attribute it to certain effects of carbon, why, we understand it. The fact of overriding importance is that at last man had at his disposal a powerful material whose potential uses were almost without limit, as future years were to show.

Gold is for the mistress — silver for the maid —
Copper for the craftsman, cunning at his trade.
"Good," said the Baron, sitting in his hall,
"But iron — cold iron — is master of them all."

Within a few centuries it was discovered that quenching carburized iron made it considerably harder still; and this knowledge was later used to advantage by the Greeks and the Romans. The inferiority of wrought iron weapons was shown as late as 220 B. C. when, in a battle between the Romans and the Gauls of Insubres during the early part of the Roman conquest, the Gallic swords of iron bent so easily that after the first thrust they had to be straightened. The Romans took advantage of this fact, and after taking the first thrust on their spears and armor, they attacked and beat the Gauls who were bent on straightening their weapons.

The discoveries made in Gerar (Palestine) show by their variety that iron was beginning to have its first real uses. Just as bronze replaced stone, so did iron begin to replace bronze, and further, many additional uses were gradually found for the metal. At the site in Gerar knives, hoes, sickles and other agricultural tools were discovered, and no doubt the Palestinians were well pleased with their new metal. But it could not be cast; and the production of molten iron, except perhaps accidentally, did not begin until over 2000 years had passed.

While the Hittites were exploiting their knowledge of ironmaking, both in trading and in waging war, the use of bronze spread through Europe as far as the Baltic. Progress in European bronze culture is shown by developments in the designs of axes and swords. The Baltic countries, in the 12th century B. C., produced gold and bronze of exceptional quality, and became wealthy lands. They traded with other parts of Europe, but presumably had no need of overseas markets. Britain, when the never very extensive bronze importations from Spain ended, made more use of her own tin deposits in Cornwall; while Ireland, which had no tin, traded in its gold; and a regular intercourse of trade between Britain and Ireland was opened up.

In an account of this kind, leaps must be made either in space or in time depending upon the review being geographical or chronological. So, jumping across to the Middle East again, we find the Hittites suppressed. History shows repeatedly that it is a country's own power which so often results in its destruction, for power breeds enemies. The Hittite empire in Asia Minor was no exception, for in 1200 B. C. it was destroyed. Egypt, weakened after its spectacular building enterprises under Rameses II, was in the power of a priesthood which, whatever effect it had upon the country's morals, had a detrimental one on its metallurgical progress. The whole of the Middle East, which up to this time had led the world, was in a state of turmoil.

China, the Baltic states and Europe, therefore, began to catch up culturally, while the rest of the civilized world marked time. Iron made a cautious

IRANIAN HEAD OF AN IBEX, IN BRONZE, 600–500 B.C. HEIGHT ABOUT 14 INCHES.

Courtesy: Metropolitan Museum of Art

entry into parts of Europe and the Baltic, and its use slowly began to supersede that of bronze. In China, the renowned Chow Dynasty had begun, and although iron was not introduced (as far as one knows) for another 400 years, Chinese civilization temporarily prospered, and the most beautiful bronzes were made for domestic use.

Chinese metal smelting was tied up so much with sex that it would confound the most Freudian analysis. Metals were male or female, and were worked and fabricated with ceremony and ritual by young virgins of both sexes and by other persons whose morals were above reproach. The limitations imposed by such a discipline need not be elaborated upon. Probably the legend of the bell-founder's daughter who threw herself into a ladle of molten metal so that her father might make a perfect bell for the emperor, hinges upon this male-female balance that had to be achieved in metallurgy, as in all the sciences that were based upon Yin-Yang. Although legends ought to have no fixed place in time, this one is said to have originated in the Han Dynasty (206 B. C.–23 A. D.). It has a pretty ending, for the young lady is supposed to have dropped her shoe as she flung herself into the metal; and afterwards, whenever the bell was rung, she called for it, "Shoe! Shoe!", or whatever that is in Chinese. Some metallurgists, dull ones, insist that the girl knew the value of phosphorus in bronze, and used herself as a human phosphorizer. This, of course, is typical of so much flat thinking. Let us keep our legends, however improbable they may seem to be.

In our enthusiasm to relate history and metallurgy, we must not lose sight of the fact that we are essentially interested in the progress of metals, their manufacture and usage. So far as we are concerned, historical events are valuable only when they bear upon metallurgical development; therefore the next few hundred years can be skimmed over lightly, for we have seen already that the use of bronze spread over most of the then known world, and that by 1000 B. C. iron was beginning to be known. Future years showed no comparable achievements, and consequently they may be dealt with briefly.

The early Iron Age in Central Europe is usually divided into two convenient phases. The first, the Hallstatt Period, is named after the center of the salt industry which existed at Hallstatt, Austria. Because the salt industry was important in an essentially agricultural community, the spread of culture quite naturally developed around it. Evidence of this was found in 1846 A. D. when a prehistoric cemetery was unearthed, with its earlier bronze weapons and later iron axes and swords. Hallstatt culture penetrated into many parts of Central Europe sooner or later, and spread northwestwards. The period ended in Central Europe about 400 B. C., when Hallstatt was no longer the focus of development and industry; and the second phase began. This was the La Tène phase, not marked by any striking metallurgical advance, but rather by a characteristic difference in style. In France the use of iron followed later; in Britain later still. Even so, iron making was fairly widespread at the time of the Roman landings in Britain in 55 B. C.

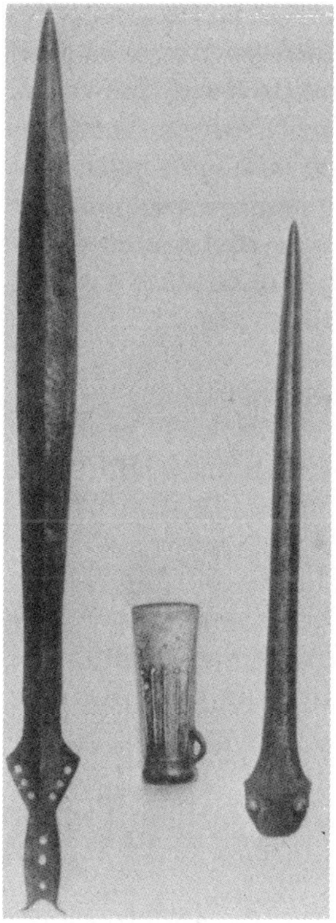

ENGLISH BRONZE-AGE SWORD, CELT AND DAGGER.

In Ashmolean Museum, courtesy of Copper and Brass Research Association

Turning eastward again, there were some significant changes. Brass (an alloy of copper and zinc) was discovered, presumably accidentally, at some time during the period under review. The exact date of the first brass to be made, like its location, seems to escape accurate definition. Dates from 1600 B.C. to 600 B.C. are cited; and Persia, China and Palestine are most commonly accorded the honor of being its birthplace. The discovery of brass must have been as fortuitous as that of bronze, for zinc ores when smelted alone produce zinc vapor, which becomes oxidized. The product is not a metal, but a scattered white powder. When, however, zinc ores and copper are smelted together the zinc vapor is absorbed by the copper, producing the alloy brass; and it was probably under these circumstances that the early articles were made. Perhaps its value was not appreciated, or the method by which it was made could not be repeated, for there is no indication that the alloy was used much or even known, apart from a few instances, until the time of the Roman Empire, when brass currency was introduced.

Roman currency was not, however, the earliest system. The first reliable one was evolved in Lydia, in the Aegean, where a standard alloy of gold and silver — electrum — was used and stamped, in about 700 B.C. This encouraged the trading that had already become well established in the Aegean region, where the valuable line of harbors in Phoenicia assumed great importance in the centuries subsequent to the downfall of the Hittites.

Phoenicia was a powerful country, and from the 12th century B.C. the Phoenicians and later their successors, the Carthaginians, controlled the mines in Spain until the time of Roman domination. They traded with Britain, and it is very possible that they bartered for Cornish tin ores, which like their own mineral deposits in Spain were to be governed by the Romans. That the Phoenicians traded with Britain for tin ores is controversial; for they, and others before them, obtained tin from what Herodotus in the 5th century B. C. was the first to call the Cassiterides, or tin islands. The whereabouts of these islands is not at all certain, and the matter is confused by the fact that the Phoenicians endeavored to mislead the Romans as to their source of metal. Hence Roman records on the subject are not reliable, and there is little other evidence of the routes which were followed, or the subsequent destination; but it seems possible that the Cassiterides were, in fact, Cornwall. Cornwall is not an island, but the geographers of that period had a carefree recklessness not found among cartographers today; and it is possible, too, that adjacent islands (now the Scilly Isles) were included among the Cassiterides, although they would probably yield little ore.

We must record in this period, too, that China began to make her own iron about 700 B. C., but her increased resourcefulness had little effect until 500 years later when the Chinese Empire dominated the East as the Roman Empire dominated the West.

Meanwhile, however, in Greece, after the defeat of the Persians in 500 B. C. there came the welcome beginnings of scientific thought.

CHAPTER 4

500 B.C. – 500 A.D.

ANCIENT GREECE was not a nation in the modern sense, but was composed of a number of city-states, some of which possessed several external colonies. Neither the geography of Greece nor the temperament of its peoples encouraged national unity; nor, perhaps, was the time sufficiently advanced for it. The mountains and the sea provided natural obstacles between scattered communities; and the independence of the Greek showed itself in his desire to rule his own city rather than to elect some one else to govern his own country. Athens, Sparta and Corinth are perhaps the most familiar names among the city-states; and colonies existed along the north coast of the Black Sea, in southern Italy and along a vast stretch of the North African coast, between Egypt and Carthage.

It was this Greece that the Persians sought to subjugate in a great empire building campaign. Under Darius I their territory included Egypt, Asia Minor and Syria in the west, and stretched eastward as far as India. Then, in 490 B. C., after the Persian fleet had conquered islands in the Aegean, the Persian armies landed in Attica where, at the battle of Marathon, they were defeated by the Athenians. Nevertheless, Athens was still threatened, and the Persian fleet, although it had sailed out of Aegean waters, remained a serious menace. Themistocles, a Greek statesman, had for many years insisted that without adequate sea power Athens could not be safe from attack, and at last his persuasive arguments were heard. But the enterprising ship building program he suggested needed vast financial resources. Fortunately, it was at this time that a rich vein of silver was struck in the mines at Laurium, near Athens. The ore was mined, the silver extracted and sold, and ships built. And in 480 B. C. they fought and destroyed the Persian fleet in the Bay of Salamis. Athens was no longer in danger, and many Greek cities that had previously been conquered were liberated. But without Themistocles' wisdom and without the silver mines at Laurium the story of Greece might have been a very different one.

The mines, leased to citizens by the state, were worked in a way that shows considerable technical development from the primitive scrapings of 1000 years earlier. In the mines at Laurium, shafts were sunk to depths of nearly 400 ft., and by cutting steps in sloping shafts access was given to the underground workings. Ventilation was helped by fires, but conditions were severe, and slave labor was used. The slaves were usually prisoners of war — frequently the inhabitants of conquered Greek cities — who were freely bought and sold in the slave market. But the conditions that were imposed upon them, although unpleasant enough, were generally not so horrible as they were under the later Roman rule. A slave might, for instance, be directed into a business house, eventually acquire property, be set free, or even be the beneficiary under his master's will. But in the mines such liberties were not common. Chains and brands were often the fate of these unfortunate people, whose lives were made more awful by the poisonous atmosphere that was produced at the mine surface by roasting the sulphide ore there preparatory to smelting it. By the light of oil lamps the slaves hacked with iron hammers for long shifts, and their produce was carried in sacks to the surface. There it was hand-sorted and crushed in crude mills or mortars; the crushed ore was subjected to a stream of water, which carried away the light, sandy material and left the valuable mineral to be roasted and smelted to produce a lead-silver alloy. This process of concentrating the mineral has remained in principle unchanged; and so has the method used by the Athenians in extracting silver from the lead by a process of "cupellation."

This involved heating the molten alloy in a shallow clay basin, known as a cupel, and blowing air over the surface. Lead oxidizes preferentially to silver and so lead oxide was produced, which was removed from the surface of the melt while a proportion was also absorbed in the material of the cupel. After sufficient oxidation, only pure metallic silver remained in the basin. The separated oxide and discarded cupels were heated with charcoal to recover the lead. Today the process is used in much the same way, except that it is made more efficient by a preliminary process which effects a concentration of the silver by simpler means. Cupellation is the same; it is only the conditions under which it is used that have altered. If gold occurred with the silver and lead in the original alloy, then the cupellation process produced the gold-silver alloy, electrum. Just when it became possible to separate gold and silver is highly controversial, but it was certainly before the Christian era, and it may have been as early as 300 B. C.

The rich resources of the mines allowed the introduction of silver coinage at Athens, and this uniform currency together with the power of the Athenian fleet brought Athens into a predominant position in trading with the rest of Greece and Asia Minor. The Athenian silver mines played an important part in history.

The Greeks were capable ironmakers, too. While the Egyptians knew the hardness conferred upon a carburized iron by plunging it hot into water, it is believed that the Greeks by 400 B. C. had learned that a subsequent tempering operation, involving heating the metal to a fairly low temperature, relieved it of its brittleness. Iron heads to lances and arrows, scythes, sickles, chain links, files, chisels, swords, razors — all these in iron (and what many would call steel) were made by the Greeks. Archimedes is said to have constructed a ship whose sides bristled with iron spikes, and which possessed other devices to discourage a boarding party. Iron was becoming important in man's life and culture.

Oh, all that is moulded of iron
Has lent to destruction and blood.
But the things that are honored of Zion
Are, most of them, made of wood.

In subsequent years the Peloponnesian Wars saw supremacy pass from the Athenians to the Spartans, and eventually to the Macedonians; but throughout these seemingly disturbed times the foundations of philosophy and scientific thought were being laid.

Of the great men that studied a wide diversity of subjects in that first rational endeavor to understand the nature of things, it is Archimedes of Syracuse who has the distinction of taking the first step towards an appreciation of metallic structure. The story of his discovery that for equal weights, different materials displace different amounts of the fluid in which they are immersed, is well known; but it is colorful enough to be retold.

King Hiero, suspecting that his goldsmiths had alloyed silver with the gold he had given them to make his crown, asked Archimedes to demonstrate whether this in fact was so. We shall never know whether an inspired guess or an involved process of logical thought led Archimedes to realize while bathing that a body displaces water according to its volume, and not according to its weight. So for equal weights, a light metal would displace more water than a heavy one; the gold-silver alloy of the crown should displace more water than pure gold of the same weight. Archimedes jumped from his bath and ran down the street shouting "*Eureka! Eureka!*"—which means "I have it! I have it!" And no doubt the citizens were inclined to believe him. The king's suspicion, by the way, was eventually proved. So the weight per unit volume, or density, of a metal was discovered to be something characteristic of the metal; and this applies to all substances, of course. A facetious mind cannot help reflecting that the course of science might have been substantially delayed had the Greeks used shower baths, rather than tubs.

If the story of Archimedes is a true one, then it is difficult to accept the Greek accounts of the touchstone, which was used to estimate the purity of gold. Touchstones, which were black pebbles of a certain kind, usually taken from the river Tmolus, had been described and their use was well-known before the time of Archimedes. On rubbing gold across them, a streak was produced whose color depended upon the purity of the metal; or if a gold-silver alloy were used, upon the amount of silver in it. According to the writings of the time, the test would allow a quantitative estimate to be made of the purity of the gold; but this is hard to believe, for not until the 15th century was the test systematized, and by a comparison with streaks made by alloys of known composition an estimate could be made.

The repercussions of Greek thought were felt after the lapse of more than 500 years, first by the Arab alchemists who, as we shall see, based their hopes of the transmutation of metals upon misapplied Greek philosophy; and later, more effectively, by the 18th and 19th century thinkers and scientists. The wisdom of the Greeks cannot be said to have had an immediate or wide effect; but the effect was there, an influence in the next two millenia.

In their writings, the Greek philosophers often mentioned metals, describing methods of mining and extraction, occasionally with amazing clarity and sometimes with annoying vagueness. In either case, their records are valuable in that they provide first-hand evidence of contemporary practices. Their descriptions include the earliest account of the production of mercury from its sulphide mineral, cinnabar. It involved rubbing cinnabar with vinegar in a brass mortar with a brass pestle — an experiment which, on account of its complete futility and absolute safety, might well be included in home

LATE BRONZE-AGE GORGET (THROAT-ARMOR).
Courtesy: Metropolitan Museum of Art

chemistry sets. That mercury was know however, is quite apparent; and, because of its liquidity at room temperature, it was not regarded as a metal. This peculiar property of mercury resulted in its having mystical properties ascribed to it — a belief that has lingered until very recently.

While Greece demonstrated its intellectual capacity during the 500-odd years before the Christian era, India, which was as far removed culturally as geographically, had begun the first real production of steel. Wootz steel, as it was to be called, was produced by a development of the carburizing process that was known to the ancient Egyptians. The Indians produced a sponge iron in a furnace that was remarkably like the earliest blast furnace (which was not to be generally introduced until the 14th century). It was about 4 ft. high, and a blast was applied at the bottom, while the charge was made at the top. But apparently the temperature reached was not sufficient to yield a liquid product. The sponge iron so made was hammered to expel the slag, and was broken up and packed tightly with dry wood chips into a clay crucible. The metal was covered with green leaves and the crucible was sealed with clay to prevent the ingress of air. A number of these crucibles was piled up and put into a stone furnace, and packed around with charcoal kept burning with the aid of an air blast. After a few hours the crucibles were taken out and cooled. The pieces of iron had absorbed carbon throughout their section, producing the alloy, steel, that had such superior properties of strength and hardness. The steel pieces were then heated and hammer-welded into bars. It was steel of this kind that later was to be used in Damascus, in making the swords for which that city became famous. These swords were made by hammer-welding together a pile of alternate bars of iron and steel, and reheating and hammering them until their thickness had been appropriately reduced. The sword that was shaped from this material had a structure consisting of alternate wafers of iron and steel, and these produced the typical damask pattern when they were suitably corroded. The corrosive agent used is not known, but it was perhaps a dilute acid, which would discolor the steel sections leaving the iron comparatively bright.

During this pre-Christian period, metallurgical activity was not confined to the heads of the Greeks and the hands of the Indians, for the small city of Rome in Italy had shaken off its imposed monarchy, and was struggling for freedom at home and an empire abroad. She attained both. By 100 B.C. the Roman Empire included all the countries at the north of the Mediterranean Sea, and during the next century or so it acquired the North African coastal countries as well as Gaul and Britain.

The Roman contribution to metallurgical progress was not so much a gain in knowledge as an improvement in organization — an efficient administration, but no back-room boys. The earliest Roman republic, near the mouth of the Tiber, had no ore deposits, and was of necessity an agricultural community. The Romans made pottery, hewed stone and mixed mortar. As their territory was extended into Italy, mines were acquired by the state, although many remained in private hands; but in either case, Rome became richer. The first mines to contribute to Roman wealth were in those parts of Italy that came under early rule. But prospecting was prohibited, and soon the existing Italian mines were closed. Perhaps this was because the Romans felt that their own soil was too sacred to be exploited; or they might have wished to conserve their resources. Further, as there was initially so great a production of gold at Aquileia, in the north, that its value was considerably diminished and prices rose alarmingly, some control had to be exercised. There was the popular belief, too, that minerals grew as crops did, and that a mine which had lain fallow would prosper on being re-opened. For reasons of this kind the Italian mines remained unworked until the conquest of Spain made gold, silver, lead and bronze so plentiful that home produce became unimportant; and the order was widely ignored.

Usually the mines that the Romans exploited throughout their empire were either leased to the highest bidder, or controlled by a state administrator, who may have worked them himself or sublet them for a rent proportional to the number of employees. There were other controls, too, and the inevitable restrictions bound up with a state enterprise of this kind produced fixed prices and stipulated markets. This was most particularly so with respect to the metal which had to be produced for the imperial mint. Chief among the contributors to the wealth of the empire was Spain, whose mines, like all Roman mines, were worked by slaves; and from the accounts one reads it appears that the slavery of imperial Rome was far more brutal than that imposed earlier by the Greeks.

But while the Athenian mines had become poorer, the mines in the south of Spain were worked to greater depths, and produced vast quantities of gold, silver, copper, lead and tin. The workings were unhealthily similar to those in Greece; and at lower levels water was frequently encountered and the mines had to be drained. This was effected by the Archimedian screw pump, which was not the new invention that its name implies. Archimedes is said to have seen it in the Nile valley in 220 B. C., where it was used for purposes of irrigation. This, however, was its first appearance in mining techniques; and it may represent the first application of a machine (unless we call picks and shovels machines) to mining operations.

But the Romans made no extraordinary metallurgical discoveries. By this time lead was beaten into sheets and pipes — an industry presumably stimulated by the Roman habit of bathing. Silver was extracted from argentiferous lead ores by the methods known to the Greeks, and already described; but we should not turn up our noses at this, for as I have said, we use the same process still. Tin was known as a metal by this time, not merely as a something which was the difference between copper and bronze; and the Romans were the first to use it as a lining for food containers. In Spain, the Romans made no improvement to mining conditions. The shafts and tunnels were small; the tools used were primitive; the conditions of working were pitiable. What the Romans did, however, was to organize their mining so that it benefited and enriched Rome. It did this, of course, at the cost of continued slavery; but in those days slavery was not a matter of social abhorrence.

The Spanish mines were leased to contractors, usually; and each mining district was under the supervision of an imperially appointed official (*procurator metallorum*). The contractor sold his metal, or used it in manufacture, according to the restrictions imposed upon him; and additionally he collected taxes from all members of the local community. On the debit side, the contractor paid his employees and was responsible for the upkeep of the public baths. This is not so peculiar as it may seem at first sight, for the local community depended upon the mine for its livelihood, and upon the baths for the only social life which a mountainous country could offer in those days.

Spain also possessed rich deposits of cinnabar (mercuric sulphide), which have been worked at Almaden, in the south, from times before the Roman occupation until the present day. The first distillation process for producing mercury, by heating cinnabar and condensing the vapor, was described about 370 B.C. Mercury's property of amalgamating with other metals was also known. Much later, at an unknown date, but probably not before 1500 A.D., the powers of amalgamation were used to extract metallic silver and gold from sandy materials. But it seems fairly conclusive that this application was not practiced by the Romans.

Just as the Romans prized Spain for its metals, so to a large extent they were prompted to occupy

Britain for its expected mineral wealth. In fact, they seem to have considerably overrated the potentialities of the country. However, they came to Rye in 44 B. C., strode out to the Severn and beyond, opening up and commandeering mines which today are pretty well worked out. They exploited with hungry efficiency the deposits which before their arrival had been worked reasonably and unhurriedly. The mineral wealth of Britain was diminished considerably during the Roman occupation, and the native Britons benefited only very indirectly. The copper mines in Cumberland, North Wales and in the Isle of Anglesey were soon put under Roman rule; the tin mines of Cornwall were unimportant in comparison with their Spanish counterparts. However, in the 3rd century A. D. the Romans occupied the extreme southwest to mine tin when the value of the pure metal for lining food containers was discovered. But, most strikingly of all, they unmercifully exploited the lead mines, for lead was particularly needed for water pipes and was even more important for the silver which it contained. The mines in Flintshire, Shropshire, Derbyshire, Yorkshire and Somerset were easy to work, for the upper deposits were still rich. All these swelled the imperial Roman purse during the 400 years of occupation.

Although the Romans contributed little to the natural progress of metallurgical knowledge, minor improvements were inevitably made. The first furnaces used in lead extraction were simple hearths built upon hillsides; but later they developed into something more furnace-like, with stone walls and hearths made of clay and clinker above ground level. Even so, the metal still had to be ladled out, for the structure was not sufficiently robust to allow a tap-hole to be spiked and prodded for the metal to run from.

To the Romans' credit, however, is the first purposeful manufacture of brass, which they used for coinage and which, for many years, had a higher value than bronze. They were aware, too, that with a lower content of zinc the alloy could be beaten into thin foil, and accordingly they used a brass containing 11–20% zinc for decorative purposes where thinness and a good color were needed. In subsequent years, brasses of this composition became known as gilding metal, or Dutch leaf, on account of their similarity in appearance to gold leaf. The brasses which the Romans used for coinage had a higher zinc content, (21–28%). But although there was an understanding of suitable compositions for appropriate purposes, there is no evidence that zinc was known as a metal, to be added to copper in varying proportions in order to make different varieties of brass. Simply, we presume, suitable combinations of copper with zinc ores were known to yield a metal which had more or less definite uses.

The Romans were responsible, too, for a wider use of existing materials. Bronze was employed not only in making statues, implements and weapons, but for furniture, cooking pots, and razors. Even an artificial leg has been unearthed, dated about 300 A. D. Lead was given an additional use, new to Europe, but first known in Egypt, as a constituent of bronze. Presumably this was only to cheapen the alloy, for the mechanical properties were bound to suffer. Although the Egyptians knew that corroded leaded bronzes produce a pleasant green patina, which looked well on statues, the Romans often used the alloy where strength was more important than appearance, adding as much as 10% of lead. One is forced to believe that the measure was an economic one; but in view of the great resources of copper and tin, this is not a very adequate explanation.

The iron production of England during the Roman occupation was not as remarkable as that of Spain, whose manufacture was indeed renowned throughout the world until the 18th century. However, iron was known and used before the Romans came to England, although only in very small quantities. Its uses were confined, and its rarity is indicated by its use in bar form for currency. During the years of Roman rule foundries and furnaces were set up, at first in the south and always in wooded areas, where charcoal could be readily available. Iron production increased to an extent where export was possible, and this trade continued until the Romans left England. It may not have been very large, for the comparative scarcity of iron ensured a high price, as the Romans were quite well aware.

The Romans did not use their wealth as coldly or efficiently as they had gained it. Much of it was spent in buying luxuries from the East. Pliny estimates that the equivalent of about £1,000,000 in coinage was sent to India and Arabia each year to pay for silks and spices and other pleasant extravagancies. "So dear do pleasures and women cost us," he said, reflecting (I need hardly say) upon the fate of the empire. The Romans worked out the readily available metal, paying it to foreign countries, even to their enemies, while internal coinage had to be debased.

The resulting economic disorder contributed to the break-up of the Roman Empire. Europe was still weak after the plague of the late 2nd century A. D.; the Roman rulers became more autocratic; and the people they governed grew more discontented. In the early 5th century the whole of Europe became prey to the barbarian invaders, and the Roman Empire, largely inspired by the lust for metals and built by their exploitation, was at an end.

CHAPTER 5

Arab Alchemy

IT IS NOT ONLY from a metallurgical viewpoint that the world was unproductive and chaotic during the period which followed the Romans' downfall. In the West, most manufacturing processes, mining — and also what, for want of a better word, we may call culture — fell to their lowest condition. The only exception seems to have been ironmaking, which still prospered. The Middle East was in a more or less continual state of war, and while in modern times it is an unhappy fact that war affords a stimulus to invention and an incentive to greater effort, in the 5th and 6th centuries A. D. this was certainly not so. China, the eastern counterpart of the Roman Empire, after a period of suppression became reunited in the Sung and Tang dynasties. It is quite possible that she was metallurgically ahead of the rest of the world; but the state of affairs there is uncertain, and had no real repercussions elsewhere.

Through this vast continent of history, we are

fortunately travelling only on the main metalled roads; and the next important system of them that we meet is that marked by the Arab conquests. By 725 A. D. the Arab empire stretched from China in the east to Spain in the west; stretched so far and so perilously that it could not be long lived. But the Arabs, despite their political shortcomings, brought about the rebirth of Greek science and philosophy, which had so long lain dormant; and through their alchemical researches they made discoveries which laid the foundations of chemistry and contributed to many other branches of science.

There is, or was, more to alchemy than the popularly supposed search for the Philosopher's Stone — that curer of ills and transmuter of metals. Alchemy implied a postulate of the unity of matter, upon which an entire philosophy could be, and was, built.

Aristotle, about 350 B. C., had propounded the theory of the Four Elements — Earth, Air, Fire and Water — material elements derived from the two pairs of abstract opposites Hot and Cold, and Wet and Dry. The theory was not entirely new in 350 B. C., for the postulate was really a crystallization of much earlier beliefs and hypotheses. Clearly, the idea of linking opposites was supported experimentally in, for instance, the transition of cold wet water to hot dry steam; and, therefore, if we are believers in the theory of the Four Elements, should it not be possible to convert any one material into any other? And surely, too, this being the case, there should be one primary element from which all others are derived. From such arguments, in the early centuries A. D. there developed the belief that there existed an agent which could effect the transmutation of one element into another, and this was named the Philosopher's Stone. Since it could "perfect" base metals into noble ones, it should also (as the Elixir of Life) be capable of righting bodily imperfections by curing all illnesses and conferring long life.

The Arabs — and in subsequent years many scientific philosophers — occupied themselves with such suppositions, straining analogy to the limit, while less scrupulous people ignored the rainbow philosophy and worked only for the promised pot of gold. If their original assumptions were wrong, although not unreasonable ones to make, the work which the Arab alchemists accomplished made a real contribution to scientific knowledge, and to the understanding of metals and their compounds. Clearly the fault of their reasoning lay primarily in equating change of state with change of inherent structure. The original Four Element theory alluded to changes of property, or quality, such as the change from solid to liquid. But the alchemists thought that the internal constitution of materials could be altered. (We now know that it can, of course; but we should not confuse the issue by dealing with the conclusions of modern science.) One has rather less patience with the 17th century alchemists than with their Arab "ancestors," for one feels that they should have learned better by their time. Indeed, Robert Boyle, in 1661, published his skepticism about the doctrine of the Four Elements and introduced the concept of an element much as we now understand it. But this was not orthodox for 100 years or so after his time.

The Arabs — to return to our chronological survey — introduced the sulphur-mercury theory, which was directly derived from the Aristotelian theory of the Four Elements. Alchemical terms were delightfully vague, and open to wide interpretation; and just as Fire was associated with combustion, so, in an abstruse way, was sulphur. Similarly, the element Water, associated with fluidity, could be identified with mercury, which was further representative of fusibility and metallicity. Thus in the 8th century, an Arabian alchemist, Abu-Musa-Jabir-ibn-Haiyan (probably Geber, and we shall call him that anyway) propounded the theory that all metals were derived from mercury and sulphur, and that, in perfection, gold would be produced by their marriage. This production, presumably, was supposed to happen in the earth, but as a result of "accidents" of impurity and so on, other minerals were unavoidably formed too. With sulphur and mercury were associated masculinity and femininity respectively (we remember that the Chinese ascribed sex to their metals). Further, the seven known metals were linked with the heavenly bodies. As Chaucer wrote later (and Coghill translated, later still):

Gold for the sun and silver for the moon,
Iron for Mars and quicksilver in tune
With Mercury, lead which prefigures Saturn
And tin for Jupiter. Copper takes the pattern
Of Venus if you please!

Geber did many useful researches. He produced

arsenic and antimony from their respective sulphides; but, of course, they were not recognized as metals in the same way that iron and copper were. Apart from the fact that the characteristics of metals were not sufficiently well known to allow it, such thinking would have been very inconvenient: after all, there would not be enough planets.

Other Arab philosophers in their common search, made similar contributions, not only to a knowledge of metals and their compounds, but to medicine, astronomy, optics and mathematics. Needless to say, their work over 700 years, to be followed by alchemical researches in other parts of the world until recent times, did not produce the Philosopher's Stone.

A colorful resumé of alchemical theory is included in "The Dialogue of Iron and the Vertues thereof" from *Joyfull Newes out of the New Founde Worlde* written by Nicholas Monardes, Physician of Seville, in 1570, and translated by John Frampton in 1577. In the course of this dialogue between "the Doctour" and "Maister Burgus," "the Doctour" says:

"All these mettalles doo proceede and are ingendred of one manner of substance original and beginning. Trueth it is, that there have beene great opinions and divers judgementes among the ancient and wise Philosophers, whereuppon the original and beginning thereof should bee and whereof they are ingendered. Some doo say that moysture congeled into a certeyne forme, is the matter of all metalls. Aristotle saith, that they are ingendred of a vapour, entred into the bowels of the earth. Damocrito sayeth, that they are made of a certeyne kinde of lime and lye. Gil, a Spanish Moor sayeth, that they are made of ashes: others say that they are made of al the Elements: some say that they are congeled of a colde cause: others, that heate doth make them thicke. The Astronomers doo attribute this matter to the superior causes, and to the knowen starres, giving to every Planet his metall. Plato, willing to reproove the one and the other, woulde that the celestiall vertue, with the earthly were the cause of his beginning and originall.

"Trimegisto sayd, that the earth was the Mother of the metals, and the heaven the Father. And Plinie sayth these wordes: The inner parte of the earth is a thing moste precious, for into it and through it doo goe, and pearce al the influences of heaven ingendering therein thinges of greate pryce, as stones and metalls: and this is done, as Calcidonio Platonico dooth say, by reason of the greate heate that is in the inner parte of it. Calisthenes understood that the form of metals were all one. Anaxagorus and Hermes sayde that the metalls had one forme in the inner part and an other in the outwarde parte, one secret and an other manifest after the manner as the lead hath within it gold, and the gold lead and so al the rest of the metals.

"There is another opinion which is that whiche is common and which for the most certeyne we do follow, which Avicen wrote in his books of Matheurous and in the bookes hee made of Alcumisto, which were confirmed by Geber and Raymond Lullis, and Arnold de Villanora, and all the rest that have treated of these matters, even unto our tyme, doo saye that the true matter of all metalls is ingendred of brimstone and quickesilver, the brimstone as the father, and the quickesilver as the mother: and the heat of the brimstone, dooth incorporate, and congele with the quicksilver in such sorte that of these twoo thinges are made the metals which are in the bowelles of the earth: and of the variation of these twoo beginninges they come to dyffer the one from the other and of the purenesse of these two beginninges, some doe come to be more excellent then other: and for this the golde is more profitable and more fayre than all other metalles, for because it is formed of his beginning cleane and pure, which was the cause and original of his perfection.

"And there were Philosophers which said that al mettals should have bin golde if it had not bin for the imperfection of the sulphur and the quicksilver. And al other metals besides the gold, they call imperfect mineralles, because they hadde not their pureness and concoction, that the gold had, with that marvellous freindshippe which nature gave unto it. And of this it commeth that the Alcumistes for to make golde, doe pretende to make cleane and purifie these twoo beginninges of the which all the metalles are made, and beeing put into their perfection the golde is made by of them which is the metall most pure and cleane of all other."

But even if the alchemists had found the way to make the "metall most pure and cleane of all other," what then? We know so well the consequences of a discovery being retained as the exclusive property of the few; and imagine so vividly the awful possibilities if it were shared. The results of anyone, or everyone, having the power of turning base metals into gold, and of achieving longevity, might be even more horrid than the results of the hydrogen bomb. One imagines that although alchemy was much more than the search for the Stone, if the implications of a possible discovery had been considered rather than the means of achieving it, the alchemists might have directed their attention to something more fruitful, or even realized the futility of their beliefs. One can appreciate the sound common-sense of the friar (of poetic fiction) who, when given the Stone,

. . . in scorn of the gift, and in rage at the giver,
He jerked it immediately into the river.

But equally well one can understand, why later, realizing just what he had done,

The friar haunted ever beside the dark stream;
The philosopher's stone was his thought and his dream;
And day after day, ever head under heels,
He dived all the time he could spare from his meals.

But all this is by the way. Fiction though the Stone may have been, the search for it was certainly real, and the alchemists' work had many valuable results and interesting effects. The results cannot be easily itemized, and no attempt will be made to do this, for in addition to the contradictory profusion of alchemical works, many factors make an exact account difficult to produce. Frequently the alchemist wrote his work under an assumed name, and the actual writer is no longer known; sometimes manuscripts that alchemists reported to have discovered, and which they ascribed to greater personalities than themselves, were, in fact, their own work;

and, perhaps most confusing of all, is the mystical jargon in which they wrote, principally in order to confuse any less learned aspirants, and to frighten them away. We should not, of course, be too shocked at this habit, for it is still practiced, though perhaps more subtly. But in alchemical writings it was carried to such extremes that it is difficult to differentiate between the factual and the allegorical. So we cannot categorize the findings of the alchemists; but we can describe the broad effect of their work.

Firstly, we are indebted to the Arabs for a new beginning in natural philosophy. Their alchemical researches led them far and wide into the fields of medicine, mathematics and astronomy, in which notable discoveries were made, and new concepts introduced. This rebirth of philosophical thought led to a logical, or quasi-logical, method of approach. Secondly, the Arabs were more than thinkers. They *did* things, introducing experimental techniques, and inventing laboratory apparatus. Geber described, presumably in the 8th century, the construction of furnaces, and also the uses of the Vase of Hermes (a sealed vessel in which many alchemical operations were carried out). It was known, too, as the Philosopher's Egg, for it was egg-shaped to symbolize fertility.

Another Arab, Rhasis, was given money to build apparatus for his researches, so that he could demonstrate the transmutation of metals to the Prince of Khorossan. (Rhasis is said to have written a book on his method of producing the Philosopher's Stone, which he presented to the Prince. When the experiments failed, the Prince threw the book at Rhasis, and the blow proved fatal—which, perhaps, shows that while a little knowledge is a dangerous thing, a whole bookful may be catastrophic.) In all the alchemical texts and illustrations of the period, it is clear that apparatus, although of a simple kind, was in use; and this represents a most important step in the progress of scientific method.

Just as Greek learning took many years to spread, so too, the results of alchemical researches were not reaped directly. They were gleaned rather than harvested. Perhaps it is because of this difficulty of pinning down any important single result of alchemical inquiry that the subject is so often dismissed as unimportant. But it did at any rate provide the vital link between Greek philosophy and the science of the 17th and 18th centuries.

CHAPTER 6

500–1500 A.D.

ALCHEMICAL THEORY did not permeate the western world for some time; and while the Moslem empire of the Arabs grew around the Mediterranean, England — invaded by the Angles, Saxons and Jutes — passed through its most obscure period. Historically the period is a vital one to Britain, for during the two centuries that followed the Romans' departure, changes took place — with little enough significance to the people that experienced them — which, among other things, were to determine very largely the character of the people, their economy and their language.

The obscurity of this period is primarily due to a deficiency in valuable contemporary documentation. Such evidence might have been literary, as it was to some little extent in Roman times, and to a greater degree in the Middle Ages. Or it might have been provided by material remains, as much of the history of the Roman Empire and before has been put together from relics which the archeologist has unearthed with the hand of care and interpreted with the eye of faith. But there is a lack of such useful evidence in early Anglo-Saxon times, for most material pertaining to this period is obtained from the sites of cemeteries.

The objects which were buried in the early Anglo-Saxon grave included articles that the occupant had made or which were thought to be valuable. But sufficient iron implements and weapons have been found in excavations to enable the archeologist to say with certainty that ironmaking was still playing a valuable part in that dark period of the 5th, 6th and 7th centuries.

Excavations have shown that iron was used for weapons (spearheads, knives and swords) and for agricultural implements (shears, scythes and sickles). But what is more important there had been no decline in manufacturing skill that prevented the production of domestic articles like chains and bowls and caldrons; nor of craftsmen's tools and personal ornaments such as brooches and buckles. The smiths of pagan Saxondom, then, did not lose their art; indeed, the view has been expressed that as a result of the departure of the Romans they were able to work with a new originality. Nor had they lost the independence that we have come to associate with smiths, for around 700 A. D. when Ecgwin, Bishop of Worcester, went to Alcester in Warwickshire to preach the doctrine of Christianity (at this time Christianity had spread pretty well throughout Britain) they demonstrated their feelings by clanging their hammers and anvils so that the bishop could not be heard above the din.

But by the late 7th century pagan practices had become rare; the habit of burying the dead with their personal belongings came to an end; Christian doctrine was generally accepted in England; and the crude civilization of the last 100 years developed into a much finer way of life. This was reflected in many directions, both in arts and in industries; and culturally Britain was further advanced than the rest of Europe. Northumbria, which included most of what is now Yorkshire, was a focus of rejuvenated industry, and produced knives of a fine quality which found their way into Europe where they were acknowledged to be superior to other types. They were made of iron that was work hardened and carburized during forging. The cementation process of steelmaking (which involves a controlled carburization, by packing pieces of iron into carbonaceous material and heating them to a temperature of 1800–2200°F.) was not introduced into England until the 17th century, although it was known in Spain and other parts of Europe far earlier. But the process of carburizing used by the makers of English knives was, in effect, fairly similar. As the knife was of thin section, carbon would be absorbed pretty well throughout; and the consistent quality of the products implies that the process was "understood" and controlled.

The fact that English knives of the 7th and 8th centuries were far-famed has an interesting significance, for evidence shows that such cutlery was made near Sheffield, which has retained its reputation as a steelmaking center.

The centers of learning were the monasteries

where, in addition, the more worldly arts of iron-making and metalworking were practiced; and in these pursuits the monks were helped by laymen who lived in their homes outside the monasteries. The metals were often used within the monasteries for ornamental and utilitarian purposes, for screens, lamps and bells; and the iron that was made was either used in the monastery, or disposed of locally, or sold to merchants from farther afield. The monks became well-known for their ironmaking and bell founding. Notable among them were Ethelwold, Bishop of Winchester, and Dunstan, Bishop of Abingdon, who in the 10th century founded bells for the Abbey at Abingdon. Each of these bishops has, apart from his greater contributions in other directions, metallurgical achievements to his credit. There is a legend of an encounter between St. Dunstan and the Devil, during which St. Dunstan seized the Devil's nose with the tongs he used at the forge. Although it is only a colorful story, it does somewhat illustrate the association between monastic life and industry (and possibly between industry and the Devil), particularly with respect to ironmaking. That monks were skilled in metalwork is substantiated by the records they left; and in the early 11th century the Abbot of Abingdon, a most capable goldsmith, was requested to make a crown for Henry IV.

After the Norman Conquest the position became very different. The newly appointed Norman abbots had no traditional skill in metalworking nor, for that matter, in most arts; what was worse, many pieces of existing craftsmanship were destroyed or looted. As a consequence, most of the necessary

works of art in the monasteries were executed by laymen who were employed for that purpose, although sometimes a man so employed became a monk, and possibly trained others within the monastery. Even so, craftsmanship was no longer so integral a part of the monastery, and it was particularly the more delicate arts, wood carving and precious metal working, that suffered.

But the learning of the monks continued, and began to spread outside the monasteries. Arab alchemy, and through the Arabs, the earlier Greek ideas, began to be known in Europe; and this was aided by the manufacture of paper, although the wider dissemination of written work was impossible until the later introduction of printing. Among the European philosophers of this period, Roger Bacon, a Franciscan monk, is outstanding for his strenuous and fearless battle against prevailing ignorance, for his amazing forecasts of scientific developments, and for his insistence on experiment. Brilliant men are necessarily born before their time, and the worth of Roger Bacon was not immediately appreciated.

The ironmaking of the monks, which did not suffer like their finer arts after the Conquest, contributed to the general expansion of the iron industry in England during the 12th century, and after. One reads of three forges given by the Norman Lord of Monmouth to the Benedictine Priory in exchange

---

BRONZE SANCTUARY KNOCKER FROM DURHAM CATHEDRAL (ENGLAND), 1133.

Courtesy: Copper and Brass Research Association

for land; of the forge in Louth Park belonging to the local monks who had permission to fell beeches and elms there for charcoal burning; and of the Abbey of Flaxley, in the Forest of Dean, where monks were allowed to take two oak trees from the forest each week as fuel for their forges.

There are two features in connection with the ironmaking of this period that are particularly important. First, that the iron was made by the same process, essentially, as the Hittites used; that is, it was forged to remove the slaggy material and hammered into shape. Second, that the furnaces and forges were heated by charcoal; and during the expansion of the iron industry this inevitably led to a severe cutting down of the forests in which the forges were operated. Eventually this produced a most critical situation in Britain.

Iron ore was taken from the earth by digging a pit about six feet in diameter. When the miner struck a seam of ore he dug away all around it, leaving a bell-shaped hole. There were none of the elaborate tunnellings that were used for mining lead. The ore was easily obtained, and when the miner had taken all the readily available ore from one bell pit he dug another. The ore was carried to the smelting furnace, which was usually situated in a wooded locality. It was more economic to carry the ore than the wood, for so much wood was needed; and as trees were consumed, so the furnaces were moved. Terrific amounts of timber must have been used, for in the Forest of Dean alone in 1282 there were 72 forges, and, one presumes, many furnaces.

Nevertheless, the supply of iron in England was

not enough to meet home demands, and foreign metal had to be imported from Spain and Germany, where techniques were similar in most respects to those at home. In fact, the type of furnace used in England, the Catalan forge (the word forge is misleading, for it was a furnace) was first used in Spain. It was a simple hearth type of furnace built in stone, into which the ore, flux and charcoal were charged on top of a wood fire, and blown by an air blast introduced by a tuyere into the charge. The slag and sponge iron collected at the bottom of the hearth, and after about six hours, the lump of iron was removed and taken to the forge. Here it was softened by a fire (encouraged by bellows) and worked until the metal was fairly free from slag and in the form of a round ball, called a half-bloom. This was taken hot to be beaten into a bar of a useful shape, and in this operation the iron had to be re-heated several times. The hammers were no longer wielded at the expense of a sweating brow and bulging biceps, but were usually actuated by a water wheel, whose projecting arms raised the hammer which then fell on the hot iron. Water wheels eventually became widespread, particularly in Sussex when that county was the center of English iron-making in the 16th century.

The furnace used by the Germans was similar to that used by the Spanish; but early in the 14th century the furnace was heightened, which made necessary a greater pressure of air through the tuyere. This was attained by actuating the bellows by a water wheel mechanism. As a result of further increases in height, the charge took progressively longer to travel through, and the time of contact between the sponge iron and charcoal was increased. Now, while the melting point of iron is 2800°F., this is progressively reduced as the iron absorbs more and more carbon; and at a composition of 4% of carbon the melting point is only about 2100°F. — a temperature that was attained in a furnace of this kind. So the product was no longer a pasty, slaggy mass, but molten.

This, then, was the first blast furnace. Over the years it has been made bigger; the number of tuyeres has been multiplied, and the blast pressure increased. Consequently its production is now vastly greater. Undoubtedly it has become the most essential of all metallurgical devices, for upon it depends the production of pig iron, from which in turn, steel is made. This was not so in the 14th century, of course. But we are now in a position to appreciate, as we can look back and see discoveries fall into their proper perspective, that this small furnace in which molten iron was first produced was the seed from which all modern industry has sprung. The production of iron by this method is the most noteworthy advance since the manufacture of bronze. As far as our civilization is concerned, it is possibly the most important discovery ever.

It is difficult to say how much this was appreciated at first, for presumably the furnace was expected to produce only more sponge iron for the forge. Instead, it yielded a liquid iron which on solidifying was hard and brittle; it could not be forged. Probably, therefore, some time elapsed before any great importance was attached to the discovery. The medieval metallurgist did not produce steel from blast furnace iron as we do today, and consequently the only value of the metal was for uses to which it could be put directly — that is, as castings. The first purposeful castings in iron are supposed to have been made in Sussex, England; but the county is only accredited with the distinction because of lack of positive evidence.

Probably more dramatic in its immediate repercussions than the production of iron in the blast furnace, was the invention of gunpowder, another discovery that cannot be given an exact date or location. Throughout the centuries clumsy weapons — the catapult, the ballista, the trebuchet and the mangonel — had been evolved; and now, concurrent with an improved skillfulness in metalworking, came gunpowder. The possibilities of this combination were obvious and the result inevitable.

Taking a rough average of all estimates, it appears that cannon were invented at the beginning of the 14th century by a German monk. That a monk was actually responsible is doubtful, and the date is equally questionable. The first use of cannon by the English is supposed to have been at the siege of Cambrai in 1339, or at Crécy in 1346.

It is indisputable, however, that the use of the cannon came rather before sound iron castings could be made, and therefore early guns were of bronze, brass or wrought iron; and they fired a stone. Within ten years of Cambrai, cannon already had an important place in the art of war waging; and their use, extended to naval warfare, was sufficient by 1345 to warrant the Keeper of the Wardrobe at the Tower of London being charged with the repair of ships' guns and the adequate provision of ammunition for them.

One cannot help wondering whether public reaction was as confused and stunned in the 14th century by the introduction of the cannon as it was in the 20th century when the first atomic bomb was dropped in Japan. For just as the atomic bomb represented the first use of nuclear energy as an explosive force, so the cannon represented the first application of the violence of a chemical reaction to war making.

The Black Death of 1348–49 had a retarding effect upon metallurgical industries in Europe. It is said to have originated in the East, and appeared in England in August 1348, spreading throughout the country so that by 1349 Ireland and Scotland were as awfully affected. Had it not been for this terrible disease, the inevitable improvements in gun making might have come earlier; for as a result of the 1,500,000 deaths (out of Britain's then population of 4,500,000) production fell, prices rose and labor was at a premium. In an endeavor to ease the situation Parliament passed the Statute of Labourers (1351) which, among other things, made it compulsory for a man to accept work at pre-Black Death wages. This law, which was re-enacted many times, was not successful despite the threatened branding of those who resisted it. Equally unsuccessful was the law of 1354 which forbade the export of iron. The general relevant effects of the Black Death, (which only put themselves right gradually) were that natural progress was delayed; that iron was doubled in price; and that steel, which was largely imported by the Hanseatic League, being very expensive already, suffered a severe loss in demand.

European steel at this time was made only by cementation. The process had developed from the primitive carburizing technique of the Egyptians, but had not altered in principle. Because of the blisters that formed on the surface of the metal during the process, the product became known as "blister steel."

Perhaps the premature stunting of the demand for steel as a result of the Black Death explains to some extent the fact that the steelmaking process was so long delayed. It is not unreasonable to suppose that, knowing as they did the superior properties that could be conferred upon iron by carburizing, and having at their disposal the means of making molten iron, the 14th century smiths might, under less trying conditions, have introduced a melting and refining technique for steelmaking. As it was, the process of "fining" cast iron (to make it more malleable) was not discovered until the late 16th century; not until 1740 was steel made by a melting process; and only in the late 19th century was a large-scale technique introduced. No doubt the long delay is principally explained by the fact that nobody had the capability or perhaps the luck to make the necessary discoveries; but in addition, there was no great demand throughout the period to provide the needed incentive. Had the Death not scourged Britain the natural increase in the use of steel would probably have continued so that its eventual production on a large scale would have become necessary at a rather earlier date.

Although the Black Death retarded the progress of the iron and steel industries, and with them, the development of the cannon, it was not very long before wrought iron guns were being made in England and used abroad. Two of the earliest wrought-iron guns, which, like the bronze and brass ones, were breech loading, were left behind by the English at Mont St. Michel, Normandy in 1424; and these are fairly typical of the cannon made in the late 14th and early 15th centuries. They weighed 5.5 tons and 3.5 tons, and were about 12 ft. long with calibers of 18 in. and 15 in. respectively. The barrel of a gun such as these was made by lashing wrought iron rods longitudinally on a mandrel, and holding them in place by heated iron hoops. When the hoops cooled, they contracted; the mandrel was removed; and the iron bars formed a barrel securely held by the hoops. Possibly molten lead was poured into the interstices, and the bars may have been roughly hammer-welded. If so, the caliber of the gun would be sufficient to allow a hammer to be used within the barrel.

By the end of the 14th century, the art of cannon making was fairly widespread. On the Continent, cast bronze and brass were used, while England employed wrought iron as well. Despite this, it is said that Continental iron was superior to English iron; and the little steel that was used in England was largely imported — principally from Spain, but also from Sweden, Norway and Germany. Frequently, the Spanish ironmaker went to English fairs where he sold his metal to the bailiff, who bought what he thought would be needed on the farm; and a

EARLY CAST-IRON BREECH-LOADING CANNON.

Courtesy: Metropolitan Museum of Art

smith employed at the farm would use it to make the necessary nails and horseshoes and tools. The great fair at Stourbridge was a market for the Spanish and the Sussex ironmakers alike. (And it is interesting to note that Stourbridge's recently antiqued namesake, old Stourbridge Village, in Massachusetts, includes two museums of early American ironware.) As a result of the insufficient amount of iron that was made in England and its important use in ordnance, it was too expensive to be used very much domestically. Jugs and pans of brass were more common than ironware, and they served a more utilitarian purpose than decorating the mantelpiece. It is popularly believed that in the reign of Edward III (1327–77) the iron pots and pans that were in his possession were ranked among the Crown Jewels.

At the beginning of the 15th century, blast furnaces were introduced into England, and this together with the steady immigration of Continental iron workers to England during the 14th, 15th and 16th centuries, led to an expansion of English ironmaking. Consequently the country became less dependent upon foreign supplies and in 1463 an Act of Parliament restricted the importation of many iron articles, including wire, the product of an industry that was later to thrive in England.

The expansion of the iron industry coincided with the increased facilities for spreading knowledge that were afforded by the introduction of Caxton's press in 1477. This eventually resulted in a wider and more rapid dissemination of knowledge than would have otherwise been possible, for until literature became widely and cheaply available the practical difficulties attendant upon studying were usually insuperable. The student had either to have the opportunity and means of obtaining manuscripts, or of travelling so that he could acquire his information at first hand.

The discerning reader — to indulge in a Victorian flattery — will have noticed that the preceding section on ironmaking has been confined to Europe, and largely to England. This is not accidental, for the art sprang up there, and apart from an unproductive period between the middle of the 16th and 17th centuries, Britain remained the greatest producer of iron until the late 1800's. Before the manufacture of cast iron is considered (a long delayed advance in view of its discovery at least 100 years before) we must leave Europe and turn westwards, where in preserving the chronology of this historical survey, we (with Columbus) discover America in 1492.

The story of the conquest of America by the Spanish is a story of the lust for gold. It is well known that Columbus was intent rather upon finding a direct route to India and China by sailing westward, than upon exploring the world for discovery's sake. When he landed on the Bahamas in 1492, the idea that he had reached India was encouraged by the gold ornaments which the natives wore, and he was told that more gold lay towards the south. He went by way of Cuba to Haiti, were he seized the gold that the natives obtained from the stream beds there. Early in the 16th century, after Columbus' death, a Spanish expedition was sent to Mexico. Off the coast it was greeted by the envoys of the Mexican chief Montezuma, who brought with them priceless presents, and a request for the expedition to leave. Tempted by the former and unperturbed by the latter, the expedition went on into Mexico. Further promises of gold had an effect quite opposite to that intended, and after a brutal and confused battle the Spanish conquered Mexico in July 1521.

At first the gold resources of the country were disappointing, for the collected treasures of Montezuma, accumulated by him and his predecessors over many years, raised the hopes of the invaders too high; but within 20 years, rich deposits of gold and silver were worked — with a percentage of the extracted wealth going to the King of Spain.

A further expedition was directed to Peru in 1531, where, aided by the circumstances of a civil war at the time, the small party finally subjugated the country. In the end, and before many years had passed, the whole of the west coast of South America, and on to include Mexico, was in Spanish hands; and the mines and deposits in the productive areas were being worked by native slave labor in a way that would have seemed brutal 2000 years before. The deposits were rich. Smelting processes were carried out in a primitive hearth without the use of bellows, but relying upon the wind, or the breath of a slave who blew through a tube. In time methods were improved, when European metallurgists were brought to the Spanish possessions; but the conditions of working remained pitiable for many years. By 1567 the amalgamation process had been introduced for the recovery of native silver. The native silver which the Conquistadores obtained was unavoidably contaminated with earthy material; and to separate these, the mixture was washed over mercury, when the silver was captured and the dirt was swilled away. The amalgam was collected and then heated to distill off the mercury, leaving behind pure silver. As the Spaniards had vast deposits of cinnabar (a sulphide mineral of mercury) in Guancavelica, the process could be economically exploited, and it led to a greatly increased production of silver from native deposits. Later the amalgamation method found wide applications in extracting gold, which behaves in the same way as silver.

Owing to unreliable evidence one cannot be certain how far metallurgical techniques had progressed before the conquest of South America. But it seems that the natives had not advanced very far in any cultures, and their extraction and use of metals were primitive; in most places they had only just begun to learn the value of bronze, and almost all of the pre-Conquest implements were made of copper, or even of gold and silver.

So, from our point of view, the exploitation of South America in the 16th century is of interest mainly because it opened up new resources which are still vast and vital. There was little contribution to the progress of metallurgical methods. But in forming our opinion of the Spaniards, we must, when we consider their conquests, weigh their bravery and exploits against their brutality; in considering their metallurgical achievements, their initial lack of progress has to be balanced against the improvements that their buying power stimulated; and the slavery that they enforced has to be set against the prosperity for which they were responsible. The final state of the balance sheet depends upon the temperament of its auditors.

Meanwhile, English ironmaking and gun founding continued, and by the end of the 15th century, cast iron cannon balls were being made and the explosive shell had been invented. The greater precision allowed by a cast iron shell, instead of the inevitably rough hewn stone shot, resulted in more accurate and forceful firing, and also put a greater strain upon the gun. By this time too, gunpowder was more certain and violent in its explosion, and a state of affairs was reached when, due to the tendency of cannons to burst, the hazards were probably as great to the firers as they were to their enemy. One can appreciate the birth of the expression "sticking to one's guns."

England quickly achieved a high reputation for her ordnance. Not only did she become independent of imports, but during the reigns of Henry VIII and Elizabeth I there was a demand for her weapons abroad. In the 1580's it was suspected that Spain was importing English guns (which, as a result of her

American acquisitions, she could certainly afford to do) in order to equip her Armada.

For a time, therefore, the export of iron and brass was forbidden; but some was still smuggled out of the country, and a number of English cannon are said to have reached Spain.

An unfortunate effect of increased ironmaking was the further destruction of England's forests. Fears began to be expressed about a future shortage of timber for domestic purposes and also for ship building; and Acts of Parliament were passed restricting the areas from which wood might be taken. The most limiting of these was the Act of 1584, which forbade the erection of new iron works in Surrey, Kent and Sussex, and ordered that iron works already in those localities were not to use timber "one foot square at the stub." Although the Acts were not strictly enforced, they served at least to check the consumption of the English forests by the iron makers. Consequently, Britain found that she had made a home market for iron that she could no longer satisfy; and she had to look abroad for supplies. In the diary of Samuel Pepys we read this entry for December 11th, 1663: "I to the coffee house, and there among others had discourse with an Iron Merchant, who tells me the great evil of discouraging our natural manufacture of England in that commodity, by suffering the Swede to bring in three times more than ever they did, and our own ironworks be lost, as almost half of them he says, are already."

That British ironmaking in the middle of the 17th century was in so poor a condition could have been no great surprise to Pepys for, as Secretary to the Navy Board, he was responsible for the purchase of timber for ship building. Therefore he must have known that the British Navy's need for timber had restricted the trade of the charcoal burners which, in turn, had reduced British iron production.

Britain, as the Iron Merchant told Pepys, looked to Sweden for iron imports. The development of the Swedish industry roughly paralleled Britain's; but Sweden, having the natural advantage of immense forests, and with no obligation to maintain or build a navy, was not compelled to cut down her production as Britain was in the 16th and 17th centuries. During this period, Swedish iron making assumed an increasing importance and was eventually responsible for about one third of the world's production – 80% of which she exported. Her exports to Britain sometimes equalled the amount that Britain herself was able to make.

Perhaps it should be pointed out here that Finland, whose iron and steel production has shown a steady increase over recent years, was, until 1809, part of the Kingdom of Sweden, and therefore contributed to Swedish output.

Of course, the production figures of these days appear small to us. We are, perhaps, so completely acclimatized to the gigantic, to 1000-ton-per-day blast furnaces, that we neglect to consider the small scale of 17th century operations and the comparatively restricted demand for iron products. Around 1700, world production of iron was between 100,000 tons and 150,000 tons per year: the iron and steel production of the world in 1956 was about 312,700,000 net tons; but, regretably, these figures do not imply that we have become any more capable of using our iron sensibly.

By 1730 Swedish production was beginning to meet serious competition from Russia; and about this time, too, the world picture of ironmaking was completely changed by the introduction of coke, which replaced charcoal. But we are jumping too far ahead, for the discovery of coke making was preceded by quite a different sort of contribution to metallurgical progress — the publication of metallurgical textbooks.

CHAPTER 7

Three 16th Century Experts

OUR HISTORY of metals, so far, has depended largely upon the findings of archeologists; or upon the writings of authors not always reliable; or inexpert, if reliable; or if expert, expert as historians rather than as metallurgists.

However, in the 16th century three books were published by three men whose metallurgical knowledge was expert. One book is *De Re Metallica,* by Georgius Agricola, published in 1556, and most capably translated into English and supplemented by valuable footnotes by Herbert Clark Hoover and Lou Henry Hoover in 1912. Another is *The Pirotechnia of Vannoccio Biringuccio,* published in 1540. The unscholarly reader is once more accommodated, since the Italian text was translated and annotated in 1942 by Cyril Stanley Smith and M. T. Gnudi. The third 16th century book is the *Autobiography* of Benevenuto Cellini; and while no one would suggest that this contains very much valuable metallurgical information, it is, at least, an interesting account of one metal worker's life.

Perhaps it would be as well to dismiss Cellini's book first. Cellini was born in Florence in 1500. His capabilities as an artist in metals were acclaimed by most people that knew his work; but no one, it seems, thought quite so much of his work as Cellini did. The life story reveals a conceit and arrogance so enormous as to be amusing; but, Cellini did have something to be conceited and arrogant about.

The books by Agricola and Biringuccio overlap in some sections but are essentially complementary. Agricola considers prospecting and surveying in addition to smelting, refining and assaying methods. While Biringuccio describes smelting, refining and assaying, he also covers casting, molding and core making and the production of certain commodities — bells and guns, for instance. If we may categorize their respective specialties, perhaps it would be true to say that Agricola was a miner and extractive metallurgist; Biringuccio was, rather, a fabrication man, who, in those times, found it necessary to make his metal before he might use it.

Since we have often followed our metal thread through a cotton-wool history and have necessarily left a number of subjects untouched for the sake of continuity, a survey of the books by Agricola and Biringuccio will allow us to fill in some of the details of metallurgical practice that have so far been omitted. However, in all fairness to Biringuccio, Agricola and their respective translators, it must be said that their texts have a completeness that I cannot hope to convey. After reading them a man might survey, prospect, mine, extract, assay, found and fabricate with the confidence of a 20th century housewife making a cake after reading, say, *Everywoman's Cook Book.* My aim is not to reproduce the cook book, but to offer a crumb of the cake.

The views of Agricola and Biringuccio were advanced for their time. Agricola debunks the divining rod (more forcibly than many of us would do today) and writes very sensibly about the value of observing natural phenomena (outcrops, vegetation, mineral springs) for effective prospecting. Both he and Bir-

DIORAMA SHOWING CELLINI WITH HIS BRONZE STATUE OF PERSEUS HOLDING THE HEAD OF MEDUSA.

inguccio were not very happy about alchemical theories, and even less about some alchemical practices. Agricola could not accept the alchemical view that all mineral deposits were formed by the interaction of sulphur and quicksilver; and, in putting forward his opinion that openings in the earth were formed by the eroding effect of subterranean waters and later filled by "juices," he annoyed not only the alchemists but those who held a literal belief in *Genesis,* which, of course, implies that rocks and minerals were formed at the same time. Biringuccio's opinions of alchemy were, roughly, that he applauded the work of the alchemists in that they were practical men doing research. ("... It gives birth every day to new and splendid effects such as the extraction of medicinal substances, colors and perfumes and an infinite number of compositions of things. This art is the origin and foundation of many other arts.") But he was impatient of those alchemists that strove to produce the Philosopher's Stone. ("The more I look into this art of theirs, so highly praised and so greatly desired by men, the more it seems a vain wish and fanciful dream that it is impossible to realize unless someone should find some angelic spirit as patron or should operate through his own divinity.") He adds that it is more commendable to "get gold from ores rather than by alchemy." However, if the metallurgist of those days did not allow himself to use some of the alchemical jargon, he would be as lost for words as a contemporary physicist forbidden to use the terms "positive" and "negative." So we should not be surprised when Biringuccio writes: "In my opinion

276 DE RE METALLICA

Hac etiam recentior & utilior eſt ſexta talis materiæ lauandæ ratio: duæ conficiuntur capſæ: in quarum utranq; aqua per canaliculū influit, è canali tranſuerſo,

THE CONCENTRATION OF TIN ORES BY WASHING, FROM **DE RE METALLICA,** AGRICOLA.

antimony [antimony sulphide] is a composition made by nature to create a metallic mineral that is overflowing with an undue proportion of hot and dry material and with its moisture poorly mixed."

There is, it seems, evidence that Agricola obtained some of his material from Biringuccio. The two men communicated with each other, and a comment of Agricola's acknowledges that Biringuccio's material did "refresh his memory." The translators of Biringuccio's book point out that this "refreshing" consisted in a literal lifting, with no acknowledgement. In fairness to Agricola it should be pointed out that he was the younger man (born in Saxony in 1494, while Biringuccio was born in Sienna in 1480), and that his interests were so diversified that he must, of necessity, have cribbed a great deal of his material from other sources. Since Agricola was a student of, and wrote texts on, philosophy, medicine and natural sciences, we should not expect him to have so great a personal knowledge of metallurgical subjects as Biringuccio who was essentially a metal worker. Whether Agricola was or was not guilty of lifting his information from elsewhere is of far less importance than the fact that he wrote down methodically and completely the information he acquired. But Biringuccio's greater familiarity with his subject is, I think, apparent in his writings.

As well as discarding alchemical hypotheses, Agricola refuted the popular opinion that ore veins in certain directions were the richest. It is not surprising that through the centuries of mining practice, ideas had sprung up about the location of rich veins — just as by Roman times the belief had grown to be generally accepted that an exhausted mine lying fallow would be rejuvenated. The essence of this current opinion was that the veins most profitably worked were those that ran east to west, and sloped upwards to the north; second best and third best veins were evolved by permutating three or four desirable directional features. And there were variations depending upon the metal, too. The reasons behind these beliefs were connected with the effect of the planets, and more particularly with the Sun, which was thought to draw out the metal if it struck a vein; therefore those that pointed or outcropped towards the south were hardly worth working. But by the middle of the 16th century these ideas were becoming shaky, and Agricola disposed of them. Similar notions, equally destroyed by subsequent experience and fact, were held about rivers and streams that carried deposits.

But Agricola's debunking did not extend to many of the superstitions of the miners, for in describing mining conditions he refers to "demons of a ferocious aspect" who "are expelled and put to flight by prayer and fasting." Some demons, called "kobalos" because of their habit of mimicking miners, were supposed to be of a gentle nature – unless they were roused. Nevertheless, German miners who sustained injuries from certain corrosive minerals held that these little men were responsible, and hence the mineral was associated with them, and was called "kobalt." This is one account, and a pretty one, of the derivation of the word "cobalt."

While some demons were relatively harmless, and others might be exorcised, others again, apparently, were always troublesome; but some of the more tangible unpleasant features of mining had begun to be eliminated. Since the time of the Athenian mines, where ventilation was provided, if at all, by fires lighted at the bottom of the shafts, more efficient devices had been introduced. Fresh air was blown into the top of the shaft by fans or by bellows actuated by man power or by the wind or a water wheel. It is significant that water wheels at the time, used for this purpose and for blast furnaces, were of the overshot type: that is, water was directed from a stream to the top blades of the wheel, which did not dip into a stream like the mill wheels of today — or perhaps one should say of a still recent yesterday.

Mining machinery was still crude, but naturally it had improved over the centuries. The Archimedian screw pump was used to drain mines in some localities, but the lift pump was more widely applied to mine drainage, either in the usual reciprocating form, or as a continuously operating machine. The motive power for these pumps, and for windlasses that drew buckets of ore from the mine, was provided by water wheels, or by men who might work a kind of treadmill, or by horses trotting a never ending circle like the grazing goat of Junior Arithmetic.

The mines were accessible by descending on a rope, or by a sloping shaft where one might (or might not) find steps or, alternatively, adopt the method described as "sitting on the dirt." In the mines, the tools used were of iron; but in design they had not altered much over 2000 years. When all is said and done, a spade is always a spade. Until real mechanization with its pneumatic drills and power-driven cutters was introduced, mining tools could not be very much changed. Some manual tasks had been eased, for the miner no longer dragged the ore along in a sack (unless conditions compelled him to do so), but used trucks that ran on wooden rails.

Although gunpowder had been invented 100 years before our stock-taking date of 1560, it was not used in German mining practice until after Agricola's death; and in England its use in this connection was delayed for a further 100 years. The method of breaking up rock was by wedges and hammers; and, also, by fire setting, which was known since the days of the ancient Egyptians. This involved lighting a fire under the rock to be broken, when the expansion of the rock caused it to crack; if it were obstinate, water would be thrown over the heated surface. The practice was a dangerous one and difficult to control; there was always the risk of a fall, and also the hazard of asphyxiation from the smoke of the fire and the fumes from the hot rock.

There had been little change by the 15th century in the methods used for breaking up the ore before it was sluiced with water to wash away the earthy material. The primitive crushers were either small hand-operated pestles and mortars or millstones that could be turned by men or horses, or by water power. Then, in the late 15th century, the stamp mill was invented, consisting of a row of vertical shafts each with a heavy ironshod block at the bottom that dropped on an iron plate. By a system of cams, each shaft was lifted and allowed to fall freely, crushing the material on the iron plate. In

LIBER NONVS. 343

Reliquorũ aũt metallorũ uenæ in fornacib. nõ excoquũtur, ſed argẽti ui
ui, ut etiã ſtibij, in ollis: plũbi cinerei, in canalibus. Sed primò dicã de argẽto
F 4 uiuo:

A BLACKSMITH'S FORGE, FROM **DE RE METALLICA**, AGRICOLA.

any crushing operation, ancient or modern, it is ideally necessary to separate each particle of earth from each particle of mineral, so that subsequent concentration (by washing, for example) can work efficiently. Depending upon the nature of the ore deposit, this may have necessitated grinding the material from the stamp mill between millstones in order to make it sufficiently fine.

The ways in which the crushed ore was treated to concentrate the mineral were many variations of a simple enough theme. Essentially, all the methods consisted of subjecting the powder to a stream of water so that the lighter, earthy material was washed away, while the heavy mineral sank and could be collected. Such a method was applied to ores of gold, silver, copper, lead and tin, and had been known for thousands of years; for gold washing dates back to about 4000 B.C. By the 16th century the water carrying the ore was directed down chutes, into tubs, through channels, and out of spouts in a way that would delight even the most surrealist plumber; but the principle was unchanged, although the method had been made more elaborate and more efficient.

Agricola described in some detail the ways in which ores should have been assayed in order to tell whether they were worth mining and extracting. Although the value of an ore deposit depends upon many local features and prevailing economic conditions, it is dependent in the first place on the amount of metal it contains. The science of assaying grew naturally from the time that a metal was obtained from its ores by anything more than an accident. By the appearance of a deposit, the miner and metal smelter might get a good idea of its value. The alchemists, with their introduction of chemical apparatus and experimental methods, made the next step; and by the 16th century there had grown the idea of assaying an ore by putting it through a small-scale process similar in most respects to the smelting operation that would be used to extract metal from it. In all probability such a method may have been employed long before, but none of the techniques was recorded until the early 16th century; and Agricola was the first to write a comprehensive scheme of assaying for gold, silver, lead, copper, tin, quicksilver and bismuth (the minerals of which were known, while its metallic properties seem to have been in some doubt). As some assays used today depend upon the same fundamental of extracting the metal from a known weight of ore, and so determining its percentage, it is not surprising that the methods that Agricola described are very similar to modern methods. Although the processes of solution and precipitation were known to the 12th and 13th century alchemists, they were not applied to assaying until some time later. They are, of course, very generally used today.

That assays should have been made by the 16th century was natural and inevitable; but that they should have been systematically described in a way that still bears critical reading is one of the first signs of metallurgy taking on a scientific aspect.

The smelting of copper, lead, tin and iron was usually carried out at the time in small blast furnaces, about 5 ft. high and 2 or 3 ft. square;

and six furnaces were built into the same structure so that they could be conveniently attended. The hearths were made of rammed clay and charcoal, and in front of the tap hole was an open hearth of the same material, from which the smelted metal was eventually collected. The bellows house was situated behind the furnace structure, and by the 16th century bellows had reached the surprising size of about 6 ft. in length. In design they were very similar to those that were used domestically in the days before fire lighters.

Chambers were sometimes built over the tops of the furnaces so that the solid material in the smoke would be deposited there for later collection, while the gases, relieved of some of their solids, passed out through a chimney at the top. Today the collection of suspended matter from smokes is a highly important economy in most works; and it is interesting that, despite the many inefficiencies of older techniques, this practice was common in Agricola's time — and it was by no means new then.

Agricola described the smelting processes and explained the ways in which gold and silver, produced by cupellation, could be separated. As we have mentioned earlier, a process was known at the beginning of the Christian era, but during the subsequent years other methods were introduced. Not only would these be difficult to explain and equally tedious to read, but in some cases it is not at all clear exactly how they operated. It is sometimes possible to deduce how they *may* have been effective; but despite Agricola's virtue in writing in such detail, it is not easy to appreciate the exact chemical conditions that prevailed while the separations were being made.

De Re Metallica contains a complete account of the method of extracting silver from copper that was to remain standard practice until 50 years ago. The process involved melting the silver-rich copper with lead, in an air-free atmosphere, and casting the resulting metal into cakes which were subsequently heated to a temperature above the melting point of lead, but below that of copper. The lead gathered the silver from the copper, and was sufficiently concentrated in silver to be run off and cupelled, while the copper was taken to be refined.

From the nature of his remarks on ironmaking and from their comparative brevity, it seems likely that Agricola was not perfectly familiar with recent developments. He referred to a furnace like the blast furnaces he had previously described, but of a greater size, and he spoke of its producing molten iron. Further, he wrote, "from this kind of [iron] ore once or twice smelted, they make iron which is suitable for reheating in the blacksmith's forge," which, as the translators point out, might well allude to a process, introduced in his time, of making a malleable iron from the brittle product of the blast furnace. This technique, known as "fining," involved melting cast iron in a hearth where it was mixed with charcoal and oxidized by an air blast and iron ore. The process is a horrible one to analyze, for while the carbon content of the cast iron was reduced by oxidation, so must carbon have diffused into the metal from the hot charcoal that surrounded it. Nevertheless — and we shall not inquire further — the carbon content of the iron was lowered, the melting point was thereby raised, the charge became pasty, and it was removed to be forged. Agricola was wisely brief in dealing with so strange a process that was, in fact, very widely used until the middle and late 19th century.

We can now take up the story from Biringuccio's book, and as appropriate introduction to it as any is the description of the founder: "The founder is always like a chimney sweep, covered with charcoal and distasteful sooty smoke, his clothing dusty and half burned by the fire, his hands and face all plastered with soft muddy earth. To this is added the fact that for this work a violent and continuous straining of all a man's strength is required, which brings great harm to his body and holds many definite dangers to his life. In addition, this art holds the mind of the artificer in suspense from regarding its outcome and keeps his spirit disturbed and almost continually anxious. For this reason they are called fanatics and are despised as fools. But, with all this, it is a profitable and skillful art and a large part is delightful."

Biringuccio gives us our first insight into methodical foundry practice. That is, insofar as foundry practice was (or, one might add, is) methodical. Good founding is not a matter of fortune, he points out, but requires skill and diligence. First, a good clay must be obtained; but an assessment of its qualities must depend upon trials. It should be "neither unctuous nor lean, neither wholly tender nor rough." It should dry without cracking, hold its shape and resist fire.

As to gating and venting molds, Biringuccio is frank: "The more vents you make in your molds and the wider the entrances, the surer you will be of a good result in your casting, if you have melted well. I neither know nor am able to say any more about this."

That a careful selection of alloys could produce a better product was emphasized: "It is possible with alloys to reduce considerably the ordinary measure hitherto used and to make guns lighter, a thing that results in easier moving of them and a great saving in expense to the patron who has them made." While Biringuccio seemed to appreciate the unpleasantness of gun making, he described it in detail, as well as the manufacture of iron cannon balls.

In a section on bell making, Biringuccio's book

REPAIRING A CRACKED BELL, FROM **PIROTECHNIA,** BIRINGUCCIO.

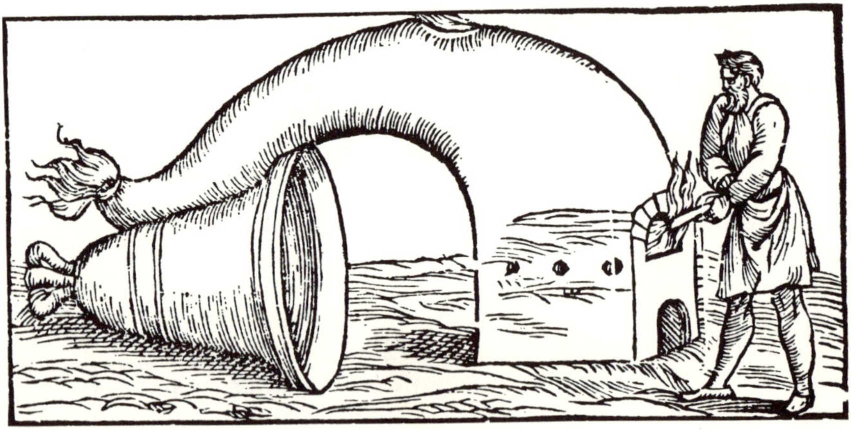

contains a surprising description of how cracked bells may be salvaged by welding them. The bell is laid in a pit and a core is fitted inside the bell. A furnace is constructed so that its flame impinges on the cracked part of the bell and, when it reaches the right temperature, melted bell metal is poured on.

A few comments of Biringuccio's on diverse subjects — and, remember they were made 400 years ago — indicate a clear perception of his subject. On alloys: "Alloy signifies nothing but the mixture of one metal with another in friendly companionship." On steelmaking: "Steel is nothing more than iron, well purified by means of art and given a more perfect elemental mixture and quality by the great decoction of the fire than it had before." On wire making: "Wire is drawn of every metal excepting tin and lead because of the need of strong bindings which must enter the fire while bound."

And we read a fascinating chapter heading — though perhaps some of its wistfulness is due to translation: "Discourse and Advice on how to operate a mint honestly and without profit."

CHAPTER 8

1500–1850 A.D.

In chapter 6 we left England's iron industry in a perilous position caused by the Acts of Parliament that severely restricted the cutting of timber for charcoal burning. Critical situations are particularly interesting to examine retrospectively; and this one is peculiarly significant, for between the end of the 16th century and the beginning of the Industrial Revolution (about 1760), ironmaking in England passed through a phase that may be experienced again, when in no more than a 100 years' time Britain will be faced with a coal famine far more critical than the timber shortage of 1580.

The use of coal as a substitute for charcoal had never been seriously considered, although its value as a fuel had been exploited since Roman times and possibly before. The objection to coal for ironmaking was based upon the volatile constituents that are evolved on heating and contaminate the iron. It was through a strange oversight that, while man knew how to drive the volatile matter from wood by burning it to charcoal (a familiar process since the very first days of smelting), the idea of applying the same technique to coal was so long unrealized. Charcoal was made by incompletely burning a pile of wood covered with earth and leaves. When the bottom of the pile was fired, some of the wood burned completely and heated the rest of the pile, which surrendered its volatile constituents, thereby producing charcoal — in an inefficient but happily rural way.

This was the sort of process that the early patentees may have had in mind at the beginning of the 17th century, when they suggested, as Simon Sturtevant did in 1612, that coal could be ridded of "those malignant properties which are averse to the nature of metallique substances." Whatever these processes were (for they were all mysteriously disguised), they failed. Then, in 1620, Lord Dudley took out a patent on behalf of his son, Dud Dudley, for "melting iron ewre and of making the same into caste workes or barrs with sea-coles or pit-coles in furnaces with bellowes." From the text of the patent rights, it seems that Dud Dudley suggested using coal and not any semi-burnt product. He tried out his scheme in some of his father's furnaces in Worcestershire, but with indifferent success. The ironmakers were averse to the scheme, and it is said that they instigated riots at his foundries and slit the bellows of Dudley's furnaces — taking the wind, as it were, out of his sales. A further setback was caused by Britain's Civil War of 1642–49, when Dudley joined the Royalists; but later he tried again, unsuccessfully, to smelt iron with coal.

Dudley has his place in history in spite of ill-success, for his efforts eventually led to more fruitful attempts to use coal in ironmaking. But it was not until about 1715 that coke was produced by Abraham Darby; and then 30 more years passed before it was regularly used in any ironworks. By this time England's iron industry had deteriorated sadly. During that Civil War many of the Sussex forges and furnaces were destroyed; and the continued fear that England's navy would suffer from a lack of timber caused ironworks in the Forest of Dean to be dismantled soon after the Restoration in 1660. From this time onwards iron production dropped progressively, until in 1740 there were only 59 furnaces (each making less than 300 tons of iron per year) in the whole of England and Wales. Britain's previous abundance of iron had made her so dependent upon it that she consumed much more per head of population than any other country in Europe; and now, to meet her demands, imports had increased enormously — to a figure about 30,000 tons in 1740.

Imported iron was largely obtained from Sweden, Russia and Norway, and, in the 18th century, from North America. The first English colonists who settled there — in Jamestown, Virginia — lost no time in exploiting the natural resources of the country, for in 1608, only a year after their arrival, they exported a few tons of iron ore to England. And in 1619 there began an ill-fated scheme to build an ironworks in Virginia. This was first delayed by the deaths of the men in charge, and later, after the Indian massacre at Falling Creek (near

Richmond) of 1622, the attempt was abandoned.

The first rating producer of iron in the United States was a plant established at Hammersmith, Massachussets. Although this ironworks suffered continual financial setbacks during most of its existence (1644–78), it was an ambitious venture. Even though it did not remain in operation for many years, it trained the men that subsequently constructed and operated other (and better fated) ironworks; and, less tangibly, it represented a remarkable industrial exploit whose eventual issue was the thriving contemporary iron and steel industry of the United States.

Strangely enough, that ironworks is, today, more substantially founded than ever before. In September 1954, after six years of reconstruction, it was opened to the public as an historic landmark. This was made possible by the co-operation of many interested people, and by the financial backing of the American Iron and Steel Institute. The following account of the history of the iron works has been summarized from the Institute's publication: *The Saugus Ironworks Restoration* (1955).

In 1641 John Winthrop sponsored the "Company of Undertakers for the Iron Works in New England" and by 1643 he had collected capital of £1000 sterling, tools and materials from British investors. The company was granted a 21-year monopoly and was set a maximum sales price of £20 per ton of bar iron. It was also obliged to attend to such nonmetallurgical endeavors as arranging religious instruction for workers and their families — a concession to employees which, so far as I know, even the most active

RECONSTRUCTION OF THE SAUGUS IRONWORKS, 1646–70.

contemporary trade union has not sought, yet which the Lukens Steel Co. of Coatesville, Pennsylvania, continues out of belief and family tradition.

In 1645 Winthrop left to construct an ironworks in Connecticut, and was replaced by Richard Leader, a metal merchant from Sussex, England. Progress was slow; but by 1648 a furnace was in operation and yielded eight tons of iron per week. However, the English investors were not altogether satisfied with the project; and, as a result of their frequent interference, Leader left the company in 1650. Responsibility for the plant was then divided between William Aubrey (sales and accounting) and John Gifford (production). Gifford, an ironfounder from the Forest of Dean in England, was, presumably, out of a job as a result of the timber restrictions imposed on English industry.

During Gifford's time, it appears that the ironworks was productive. A labor force was drawn from Scotland (the men were sold by Cromwell whose forces had taken them prisoner), and were cared for by Gifford with what appears to have been an unwarranted generosity. Gifford introduced extensive developments that led to financial problems; and in 1633 a series of legal battles began between the interested parties. One of the outcomes was that Gifford was imprisoned; but he returned to the ironworks in 1658 with hopes of restoring the plant and, probably, of vindicating himself against his old enemies.

When Gifford returned, the company was again under new management, and between 1658 and 1663 production figures (but, regrettably, not the profits) were good. Production petered out in 1665; by 1678 the plant was abandoned.

Other plants had been built by this date, however (including Winthrop's in Connecticut), and they were producing enough iron to export to England. The American iron industry has not looked back since.

It was not until the middle of the 18th century, however, that iron production in Britain showed any improvement; and, of course, this new life was to be short-lived. The use of coke in ironmaking was now accepted, and new furnaces and forges were established — not in the forests, but on the coalfields. As a result of the good quality of the coal, and the increased demand for iron, the industry expanded rapidly, and during the next 100 years Britain was the greatest iron producer in the world.

The important aspects of this 200 years of ironmaking are then: first, that the manufacture of coke overcame the problem of timber shortage (and without coke making it is difficult to believe that industrialization could have occurred); second, that North America has already and so early crept into the picture.

But in presenting this account of the fall and recovery of British ironmaking and the birth of American industry, we have skimmed rapidly a period of two centuries, which included many other important advances that were to contribute during the next 100 years or so to the spectacular progress of European industry. From this point our account becomes geographically centered around western Europe. As far as metallurgical advance is concerned, the East, which last shone brightest during the Arab conquests, has long passed its heyday; the North American continent has not yet predominated the scene. Western Europe, led by Britain, is to enjoy a period of a few hundred years when it leads the rest of the world in production, as well as in the research and discovery, industrial and more purely scientific, that made its industries thrive.

One achievement was the establishment of the steelmaking process that was to be the forerunner of today's techniques. This was the crucible process, introduced by Benjamin Huntsman, a Doncaster clock maker, about 1740; and, although it was basically similar to the Indian method of making Wootz steel (and therefore, like almost everything else, cannot be said to be entirely new), this was the first steel melting process sufficiently reliable to be accepted and used in many parts of the world.

In making his clock springs, Huntsman was obliged to use "shear steel," which was produced by forging bars of "blister" steel together until the irregular carbon content across the section had been fairly well evened out. Despite these efforts, however, the steel was not sufficiently homogenous for Huntsman's purposes, and he re-conceived the idea of melting and casting the metal to produce a uniform product. To keep his secret he moved to the small village of Handsworth, near Sheffield, and conducted his researches there. His greatest obstacle was to find a sufficiently refractory material for making his melting crucibles.

At length this difficulty was overcome, and by 1770 the steel made in Huntsman's factory near Sheffield was acknowledged to be far superior to blister steel. The process involved melting blister steel with a flux in a clay crucible (about 1 ft. high) which was surrounded by an air-blown coke fire to produce the necessary high temperature. After five hours or so, the metal was cast into ingots and subsequently forged. It was hard and strong, and would take a fine edge. It therefore came to be used for making knives, blades and razors; and it is still acknowledged by a few steel users to be unsurpassed for these purposes. Sheffield, already known for its steel, became famous, and many legends grew around it. One of these was that the water in Sheffield was superbly efficient for quenching purposes, and that this was in some way responsible for the good properties of Sheffield steel. Until the late 19th century, water was sometimes shipped from Sheffield to steelmakers in other parts of the world, who attached an almost religious belief to its powers.

Huntsman's steel was very expensive, and its use was necessarily limited to tools and special machine parts. It had no constructional applications; no one thought seriously of using steel in a big way. But the process was more than a method of making high-quality steel, for it showed the way to the more economic processes introduced by Bessemer and Siemens in later years.

Crucible steel was first made in the United States of America about 1812. The early experiments were not successful; but when a good quality clay was found for making the crucibles, the situation appeared to be more hopeful. However, it was not until two brothers, Garrard, emigrated from England to Ohio, where they established the Cincinnati Steel Works, that a good grade of crucible steel was regularly made. This was in 1832. The financial disturbances of 1837 were responsible for closing the plant; and after that date, for many years, imported steel undersold the American product. Crucible steelmaking reached an all-time high in the United States at the surprisingly late date of 1916 when 130,000 tons were made; but this represented only 0.3% of the total steel production of that year.

This account of the development of ironmaking and the introduction of steelmaking in the 17th and 18th centuries should illustrate how western Europe only gradually became ready for what is so inappropriately called the Industrial Revolution – inappropriate, because the process of industrialization was essentially evolutionary.

An industry that was later to flourish with steel manufacture was that of making tin-plate. The date at which tinning was first practiced is as obscure as the date at which tin was first made. However, in 1620 the art was introduced into Saxony from Bohemia where it had sprung up, and according to Andrew Yarborough, it was brought by him to England in 1670, when a plant was opened in Monmouthshire, Wales, that was not to become firmly established until 1720. (Perhaps it should be made clear here that tin plating is not an electrical process like chromium plating is, for example. It essentially involves dipping sheet iron or steel into molten tin, which adheres to the sheet, solidifying to form a protective and fairly noncorrosive film.) During the late 18th century the canned food industry developed, and contributed very largely to the expansion of the world's tin trade. Today it consumes about one half of the world's tin production.

But the manufacture of tin-plate was only one of the industries that grew in Wales. The wire drawing trade was expanding too; and this was largely made possible by the erection of the first water-powered works at Tintern, Monmouthshire, in 1567. In 1566, rich copper deposits were discovered in Cumberland, and zinc was found in Somerset. The Society of the Mineral and Battery Works was granted the monopoly of manufacturing brass and iron wire (1568), and by 1605 many different grades of wire were made by them or by their lessees. Most of the wire was used for wool carding — for the wool industry was flourishing. In 1630 a Royal Proclamation prohibited the importation of any foreign wool cards, and also the trimming and resale of old ones. The remainder of the brass wire was used for making pins. It is a remarkable fact that, although pins were made by hand, a good worker produced up to 24,000 pins each day.

The monopoly, however, eventually resulted in a lack of development in the industry, and this led to an Act of Parliament (1689) which removed restrictions on the mining and use of copper. Authorities have not always acted so commendably. The increased accessibility of copper resulted in the erection of many battery works — so called because of the way the metal was hammered into sheets. These increased the exploitation of and found more uses for copper, and its alloys bronze and brass.

Most of the copper ores mined in England, as well as the imported ones from Spain, were smelted in South Wales at Neath, and later at Swansea, where, by the middle of the 18th century, coal was used instead of charcoal. It was in Wales that the reverberatory furnace was developed. Its essential difference from the blast furnace is that the fuel and ore are not in contact with each other. The early reverberatory furnace consisted of a shallow rectangular hearth over which hot gases from a coal fire were drawn by a good draft, and the heat was reflected on the charge by a low roof. It was particularly suited to the smelting of sulphide ores (where treatment was a complex operation), because it was easy to control. Whereas the descending charge of a shaft furnace could not be readily adjusted, the reverberatory furnace allowed conditions of temperature and atmosphere to be altered, and each charge could be dealt with separately. Although the copper smelting process has changed very considerably since those times, the reverberatory furnace is still quoted in textbooks as being the better process.

Britain became by far the largest producer of copper, and during the 18th century found a greater use for the metal in the alloy with zinc. This alloy, brass, was first made by heating copper surrounded by zinc ore and charcoal; but since that time it had been discovered (when, it is unknown) that brass could be produced by melting metallic zinc and metallic copper together. This, of course, necessitated the production of metallic zinc, a process that Europe learned from the Chinese about 1730. Zinc smelting is complicated by the fact that the temperature at which reduction takes place is above the boiling point of the metal. Zinc vapor is therefore produced, which is so readily oxidized that the production of pure metal is difficult. The process that became established in England about 1740, involved heating the carbonate ore of zinc with charcoal in a sealed clay pot with a tube inserted through its base. The vapor did not oxidize, for the atmosphere in the pot was mainly of carbon monoxide, and metallic zinc trickled down the tube. By 1760, a process had been patented in England for roasting the sulphide ore of zinc to convert it to oxide, which was subsequently reduced to metal in the same way. The distillation process developed in Silesia and in Belgium, and eventually consisted of heating clay retorts containing the charge, while the metal was collected in clay condensers that were luted on the mouths of the retorts. For some time European use of zinc was limited, and the metal was of greatest value as an export; but gradually it found applications, particularly in brass, which is fairly strong easily cast, and will withstand corrosion. So it became used in engine parts, valves, pumps, and in the machinery that was rapidly finding use in expanding industries.

Just as zinc production in England depended upon an art learned abroad, so the development of England's lead industry during this period was largely due to foreign skill, for Germans had been brought to the country during the reign of Queen Elizabeth I (1558–1603) to improve the techniques and production of that industry.

The Crown also encouraged mining enterprises: restrictions on felling timber were more completely removed; and the burdens of taxation were eased. As a result, Derbyshire and Cumberland became fairly productive regions for a time, in copper as well as in lead. But many of the ventures petered out by the middle of the 17th century, due to a number of factors to which Britain's Civil War was contributory. After a period of inactivity, lead smelting was revived; and from about 1750 for 100 years, England experienced her brightest period in lead production.

The reverberatory furnace was introduced into Derbyshire lead smelting in the middle of the 18th century, and replaced the "ore hearths" that had previously been used there. The ore hearth was essentially a small shaft furnace with an opening

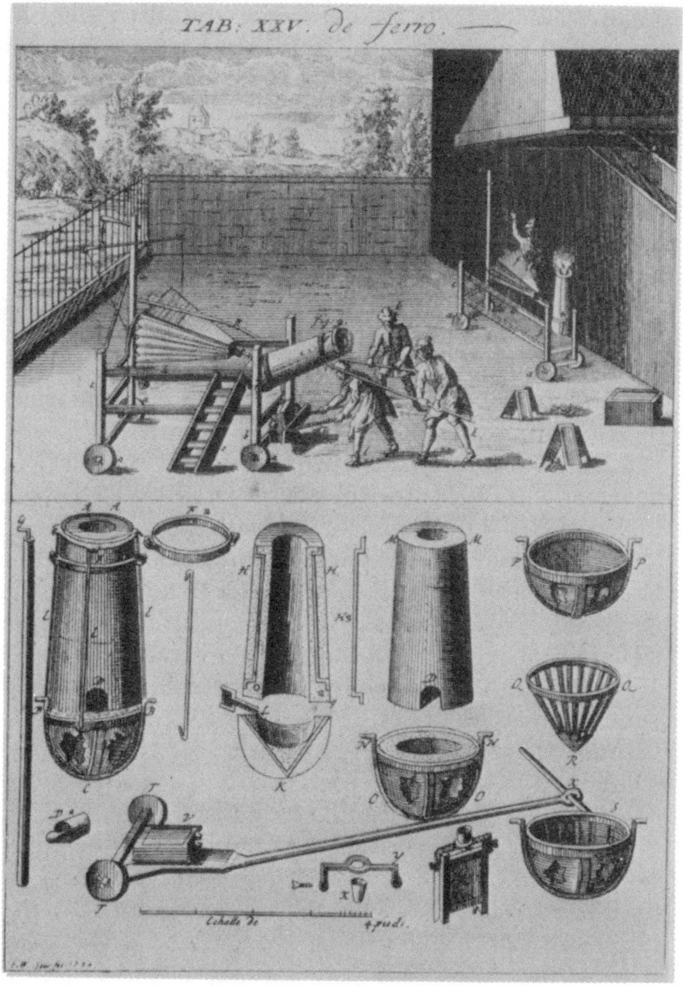

AN 18TH CENTURY PRECURSOR OF THE BESSEMER CONVERTER.

on one side from which some of the charge could be pulled out onto an iron plate where it was worked to remove slag and to encourage the oxidation of the sulphide. As the ore hearth required the use of charcoal, the coal-fired reverberatory furnace was a popular innovation; but in some localities the ore hearth remained in use, and with the advent of coke it regained a little popularity. It still enjoys a restricted use in some of the southern United States. English lead production was already great and increasing; and from the early 17th century, metal and ore were exported to the Continent where the Dutch (who were more skilled in these matters than the British) extracted silver from them.

Progress was not limited to industrial production, however. In 1751, nickel was isolated by Cronstedt in Sweden. The metal had been unwittingly used from very early times in iron of meteoric origin, but Cronstedt was the first to produce it as a metallic element. Like several other metals, the possible uses for nickel were not realized for many years.

Although compounds of arsenic and cobalt had been known for a long time, it was not until the turn of the 17th century that the existence of the elements was appreciated. Arsenic was prepared by Schröder in 1694; Brandt recognized cobalt as an element in 1735. Antimony and bismuth, mentioned by Agricola in *De Re Metallica,* had been investigated and their properties determined by the middle of the 18th century.

In 1754, Marggraff (a German chemist, better known for his detection of sugar in beet) recognized that alumina (aluminum oxide) was a distinct mineral; and some authorities hold that he identified the metal aluminum. This is, perhaps, doubtful, for metallic aluminum was first produced by Wöhler 70 years later, in 1828; and aluminum production on a commercial scale was not made an economic proposition until 1887.

These discoveries and investigations had direct effects that are readily apparent; but it is more difficult to analyze the results that other scientific research of the period had on metallurgical practice. The 200 years from 1550 were rich with scientists and philosophers — Newton, Galileo, Boyle, Huygens, Gilbert and many others — who contributed to the natural sciences. Since the days of the early alchemists (the later ones had become so obsessed with hopes of finding the Philosopher's Stone that their factual observations and logical argument had suffered) this was the most fruitful era in scientific and philosophic thought.

Perhaps, in a metallurgical context, Boyle should be specially mentioned, for if he was the father of modern chemistry, then he was certainly the unsuspecting godfather of the metallurgical sciences. Although many of the scientists and thinkers of the time were able to take the Greek philosophies as a starting point, Boyle fortunately rejected the Four Element theory, which had grown so out of hand, and differentiated between elements, compounds and mixtures. Confusion between elements and compounds was apparent in the comparatively lucid *De Re Metallica;* a distinction between them was vital before there could be an understanding of the chemical reactions during metallurgical processes.

This summarizes the state of affairs before the Industrial Revolution overtook Britain. Then, the inventions of Hargreaves, Arkwright, Crompton and others mechanized the textile business; and spinners and weavers were forced to abandon home industry to contribute to the output of the factory where the new machinery had been installed. At first the mills were water driven, and they were built in the northern counties of England where such power was available; but before long the steam engine was introduced, and the industries, already firmly established and fortunately placed with respect to coal fields, remained.

The first application of a steam engine, so primitive that we would probably not recognize it as such, was in draining coal mines in the early 17th century. Through the work of Savery, Newcomen, Boulton, Watt and Murdoch (not to mention Polzunov, the sole Russian inventor) the machine was developed and improved; and by the end of the 18th century it had been applied to mining for pumping machines, to smelting for actuating the blowers, and to driving the looms and other equipment of the textile industry. The engine, itself depending largely upon iron components, increased iron production in pit and smelter, as well as increasing the iron consumption in the factories.

In this way, and in the mechanization of many industries transplanted from the home, the demand for iron grew; and while cast iron was suitable for many purposes, its brittleness prevented its really extensive use. The process introduced during the 16th century for "fining" cast iron has been mentioned in Chapter 7; its inherent disadvantage lay in the conflict between the air blast decarburizing the iron, and the charcoal in intimate contact recarburizing it. Understandably, it was necessary to use a powerful air blast to keep things going the right way, and this in turn burned up a large amount of charcoal. To add to this metallurgical struggle the hearth was often water cooled, and water was thrown on the charcoal to prevent its burning too quickly. All these factors make up so difficult a balance sheet to explain away, that it is relieving to find that in 1784 Henry Cort patented an alternative process. Its essential difference was that contact between fuel and iron was avoided, for Cort successfully introduced the reverberatory furnace into iron metallurgy. Probably the earliest application of the reverberatory furnace to ironmaking was in 1766 by the Cranege brothers, but Cort effected so many improvements that the process became no longer a curious exception, but an accepted practice. The possibility of carburization was eliminated, and with it the extensive blowing machinery that had been necessary.

In Cort's process — the "puddling process" as it came to be called — the hot gases from a coal fire were drawn over pig iron that was charged on the furnace hearth. The charge melted and was oxidized by additions of iron oxide (in the form of iron ore or mill scale). These additions also contributed to the bulk of iron that was produced. As oxidation progressed, the charge was stirred by the furnacemen using long iron rabbles, and as it became pasty it was worked into balls weighing about

80 lbs. each, which were subsequently forged into salable shapes.

Cort's process yielded the metal that we have come to accept as wrought iron. It was not only low in carbon, but in silicon, manganese and phosphorus — elements that are present in pig iron and contribute to its brittleness. This new material unavoidably contained slag, just as the earliest smelted iron did; but as this was strung out into fibers by the forging operation, it was not harmful. On account of its malleability, wrought iron found many applications, and until the late 19th century it was the most valuable constructional metal.

Another equally important development due to Cort was the use of grooved rolls to make wrought iron bars (patented in 1783). Rolling mills, like reverberatory furnaces, were known before Cort applied them to the iron industry, and indeed, it is questioned whether he was, in fact, the first to introduce them to iron rolling. Whether he did or not, he did not benefit by his ingenuity.

When he contracted to make rolled iron for the Royal Navy, Cort invested his entire fortune in expanding and developing his process, and went into partnership with Samuel Jellicoe, whose father, the Deputy Paymaster to the Navy, contributed over £50,000, and was to receive half the profits that accrued to the partners. In those days, apparently, such a transaction was more openly made than in these; and in this case the British Government itself sanctioned the partnership. But even contemporary opinion did not allow Mr. Jellicoe Sr. to borrow £27,000 of his capital from the Navy accounts; and on his death in 1789, when this was discovered, Cort's patent rights were confiscated and Cort was removed from his rolling mill. Strangely enough, Samuel Jellicoe was put in charge. This action led both to the prosperity of British ironmasters (for they no longer had royalties to pay) and to Cort's ruin. But Cort's contribution to British industry, unrewarded as it was, resulted in a vastly increased iron production — from about 18,000 tons a year in the middle of the 18th century to 4,000,000 tons in 1820.

While the causes of the Industrial Revolution in Britain are as complex as the historian cares to make them, there could have been no subsequent effects had it not been for the manufacture and use of wrought iron. Britain showed her metallurgical ability, too, in developments in extracting metals, in the exploitation of mineral wealth, the manufacture of steel and a growth of industries that depended upon basic metal making processes. These, although they did not precipitate a revolution, at any rate induced the new era: Britain was the first country to experience the glories and infamies of industrial prosperity.

CHAPTER 9

New Enterprises, New Regions

THE YEARS that we have covered so far – from an uncertain date thousands of years ago to the beginning of Europe's industrial era – complete what may be called metallurgical ancient history. What follows is a profusion of developments, the main features of which will be analyzed in this and the next chapters.

Modern techniques and inventions are almost always founded upon the more cumbersome practices of years ago, and the continunity that exists between the old and the new ought to make itself apparent. But to bridge the gap that is so often supposed to exist between "then" and "now," nothing could be more appropriate, perhaps, than a chapter on the "new" countries – North America, Australia and South Africa – whose metallurgical history just extends into the "ancient" past as we have defined it.

Although the development of American industry is mainly concentrated into the years 1850–1950, many of the mining exploits had earlier beginnings.

The early colonists were just as keen to realize their new country's wealth as the Spanish Conquistadores had been in the south; and, although their conquests did not always involve the massacres and brutality that is associated with the Spanish adventures, the Indians who were displaced from parts of North America were often far from amicably disposed towards the immigrants. It is fairly certain that the natives of the country had neither the desire nor the ability to utilize North America's mineral deposits. They used native metals, copper and gold; but they neither dug ores nor smelted them. It is almost inconceivable that only 200 years ago so much of the world was so happily uncivilized.

Virginia and Massachusetts were explored by Europeans pretty well as soon as they arrived, and in 1606 a charter by James I stipulated that one fifth of the precious metals and one fifteenth of the copper that was discovered by the British colonists should be surrendered to the Crown. With 3000 miles between colonists and Crown, it may be surmized that James I did not receive his full dues.

As we recorded in Chapter 8, the first production of iron was in Massachusetts, where in 1643 a company was founded, and operated intermittently. Other companies were rather slow to be formed, but (doubtless on account of the easily accessible metal deposits, particularly gold) no complicated machinery, financial or mechanical, had to be introduced until much bigger business was contemplated. One presumes that the prospector was able to barter direct with traders who visited the country to deal in furs; and in the Mississippi country, for example, fur trading and metal trading were carried on simultaneously by the same men. The native Indians had learned from the French pioneers the art of smelting lead ores, and they disposed of their metal to American and Canadian traders. The Mississippi area was, in fact, one of the first places to be mined for lead on a large scale, and from 1700 until the present day the states of Illinois, Missouri and Iowa have contributed handsomely to American lead output. The other lead areas (in Utah and Idaho in the United States, and in British Columbia, Canada) were not in production until some time later.

Things are not what they were, of course. The typical adventurer of a century or so ago was prospector, miner and metallurgist, and had a combination of brain, brawn and business acumen that makes us respect him. But even then he could not be expected to be very competent in all of his duties. He had to be a good prospector to find ore deposits; he had to be a sound businessman to keep his head above water; but, although his metallurgical knowledge had to be real, there was little chance of its being elaborate; and frequently, as a result of its insufficiency, the early enterprises were erratic in getting under way.

The story of the American pioneers and prospectors of the 18th and 19th centuries reads like a Hollywood film script, and it is reassuring to find that the old-time prospector film plot has some essence of factual truth. One reads for example, in T. A. Rickard's *A History of American Mining,* of Noah Kellogg, who in 1885 prospected for gold in

Idaho, and found a vast ore body of lead. (An unkind legend has it that his donkey made the discovery.) Only two years afterwards one of the greatest lead companies was incorporated as a result of his find; and in 1892 there was a bitter disagreement between mine owners and employees, which resulted in a battle between the guards and the miners. In the course of the fight, a freight car was filled with gunpowder, fused and set off down the mine railway track. Next an ultimatum was delivered that if non-union miners were not fired immediately, then the premises would be. And, of course, as in a film story, it all ended more or less happily.

More spectacular still is the story of the great California Gold Rush. The facts are — or appear to be — that on January 23rd, 1848, James Marshall, a carpenter at Coloma, noticed flakes of gold on the bed of a mill stream that had run dry. The avalanche of prospectors was set off by the laborers at Marshall's saw mill and the neighboring ranchers, who gave up their jobs to go gold washing in the streams of the locality. Farmers, builders, almost everyone left his work to make his fortune, and within a few months the hunt had reached such fever pitch that even the San Francisco paper, *The Californian,* suspended publication, and its staff went from one occupation of the Devil to another. All California, most of America, and part of the outside world joined in. In the year 1848 alone, the thousands of prospectors and gold washers in the Sierra Nevada country were rewarded with $10,000,000 worth of gold.

By the end of the 19th century this new enthusiasm in a new country had led to high productivity. The gold riches of California were only part of the country's vast resources. The Mississippi region, Utah and Idaho were producing increasing quantities of lead; copper was mined in the Lake Superior district, Montana and Arizona; and iron was being discovered and worked in so many localities that it would be tedious to list them.

Although the United States' mineral industry has a late enough beginning, most of Canada's deposits were not worked until even more recently. Apart from iron ore, which was mined prior to the British conquest of Canada in 1763, minerals do not play a part in Canada's economic history until the late 19th century; and they assert themselves most markedly over the past 50 years. The vast resources of the Canadian Shield were first effectively tapped at Sudbury, Ontario, where copper deposits were revealed in 1883, during the construction of the Canadian Pacific Railway. These deposits proved to be a disappointment to the original prospectors, because the nickel present could not be effectively separated, and was without real commercial value at the time. Within a few years, however, the importance of nickel in the manufacture of alloy steels was appreciated, and a technique of separation had been discovered by Robert Thompson. Since then, although the copper production of the area has been considerable, and large amounts of silver, gold and precious metals have been recovered as a very profitable sideline, Sudbury has been most important for its nickel, and today accounts for roughly three quarters of the world's output. During the early part of the 20th century, Ontario revealed her resources of cobalt, silver and gold, while British Columbia's lead-zinc deposits at Kimberley were successfully smelted after an unsuccessful beginning; and although earlier copper deposits in British Columbia have been depleted, new reserves are being developed. Canada's wealth has been swelled further during the last 50 years by the exploitation of copper in Quebec and in Manitoba, where rich zinc deposits also occur. The Canadian northlands have recently moved into prominence as uranium mining areas; and explorations indicate that other metals may be worth mining, despite the difficulties indigenous to these latitudes. Another metal of recent interest, titanium, is found in Quebec.

The huge metal reserves in Canada, together with widely scattered power sources — coal, oil, gas and hydro — are beginning to change the character of many areas whose economy, until a few years ago, depended entirely upon farming.

So, in little more than a century the North American Continent has discovered its riches; and, with the experience of 6000 years behind them, its inhabitants could hardly have done other than build and sustain a prosperous civilization.

The early metal production of the North American Continent has a definite place in a history of metals, for the mining ventures of the 19th century were in many cases very soon followed by extraction and fabrication processes. These resulted in a production of most commercially valuable metals (except tin), which, by 1890 had exceeded that of Great Britain, and had strongly affected world economics.

South Africa and Australia, on the other hand, did not enter the fields of extraction metallurgy so immediately, and never with such impact. Their 19th century histories are closely bound up with discoveries of gold, and an account of the rushes that followed (with all their economic and political repercussions) is more appropriate to a history of mining than of metallurgy. The interested reader should refer to T. A. Rickard's *Man and Metals.* But it is impossible in a book of this kind to draw a sharp line of demarcation between mining and metallurgy, and there is reason enough to mention the barest features of the Australian and South African enterprises.

Although the existence of gold in Australia was known before 1851, Edward Hargraves made the first fruitful discovery in February of that year in New South Wales. The subsequent rush seems to have been fairly well controlled by the governor — unlike the situation in Victoria, where the discovery of gold at Ballarat in September 1851 was followed by a confused stampede. In 1852, Victoria and New South Wales together produced gold to the value of £15,000,000. The increased gold in circulation did not have the disastrous effects that many economists prophesied, but rather stimulated world trade. As a consequence of gold prospecting in Australia, the country became pretty well explored; and before long, the lead-zinc-silver deposits of Broken Hill and the tin ores of Tasmania were exploited. The population increased; agriculture thrived to support it; and the country so recently colonized prospered.

In South Africa, this century has seen developments in the copper production of Northern Rhodesia and the Belgian Congo, which together produce one fifth of the world's total. A little older established is South Africa's gold industry, which today enjoys being responsible for almost one half of the world's output.

But the search for gold has never been without its seamier side. In North and South America the resistance offered by the natives led to bloodshed. Even in Australia there were riots among the prospectors. And although the earlier "gold rushes" of the alchemists were not marked by obvious brutality, one cannot help reflecting that lives were sometimes unnecessarily wasted. In South Africa the results of gold mining (and diamond mining) were no less unfortunate; and they are, once in a while, brought to mind by an occasional encounter with an "Old Soldier."

CHAPTER 10

1855–1957 A.D.: Aluminum and Electricity

IN THE BRONZE AGE we saw the slow introduction of iron into the lives and habits of the ancient peoples. And now, at the beginning of the 19th century, with iron so well accepted a constructional material as to have been at the foundations of industrial development, the new metal aluminum makes its important entry into the culture of a more advanced civilization.

Even the most enthusiastic salesman of aluminum would hesitate to say that this metal will ever supersede iron and steel in the same way that iron superseded bronze. But on account of its lightness, its strength when suitably alloyed, and its pleasing appearance and resistance to corrosion, the uses of aluminum are steadily increasing; new applications are still being discovered; and it would be foolish to believe that the applications of aluminum have yet been fully exploited.

In 1845 Wöhler was able to prepare enough aluminum to determine its density, and to establish

ALUMINUM PRODUCTION CELLS, FORMING A "POTLINE" AT ISLE MALIGNE, QUEBEC, CANADA.

Courtesy: Aluminium Limited

that it was ductile, stable and easily melted. In 1854, Deville, a Frenchman, was able to produce larger quantities of aluminum, and public opinion was at once aroused. (The Emperor Napoleon III, for instance, wondered if it might be used for helmets and armor.) By 1857, aluminum was being sold at about $20 a pound – a 20th of its price five years earlier.

Aluminum is not an easy metal to produce, for its common ore, bauxite (aluminum oxide), is very stable and cannot be smelted in any ordinary way. Deville's process was quite an elaborate one which depended upon converting aluminum oxide to aluminum chloride, and reducing the chloride with metallic sodium. (It is interesting to notice in passing that zirconium and titanium, metals even more reactive than aluminum, are presently produced by a similar sort of process.) Then, in 1886, the Cowles brothers introduced a technique which involved passing an electric current through an intimate mixture of carbon and bauxite. At the very high temperature attained the oxide breaks down, but only, unfortunately, to form aluminum carbide. The Cowles brothers ingeniously overcame this difficulty by allowing the aluminum to alloy with molten iron as it was formed; but, of course, to their disadvantage they produced an iron-aluminum alloy. The process was not long lived, for in 1887 Hall in the United States and Heroult in France independently devised the method of making aluminum that is used today.

It is significant that all these techniques depended in some way upon the electric current. In Deville's method, metallic sodium for the reduction operation was produced by an electrical method; the Cowles process involved electrical heating; and the Hall-Heroult method depends upon the phenomenon of electrolysis.

The introduction of electrical methods into metallurgical practice represents the most revolutionary change — not necessarily the most important — in an entire history of development. The use of electricity for power and lighting in all industries was an innovation of the same magnitude as the invention of the steam engine. But in the metallurgical industry there were even further reaching effects, for the harnessing of electricity brought with it methods of producing and refining metals that bore no relationship to the trends of the previous 6000 years.

These new methods depend upon the process of electroplating, which is so inextricably bound up with 19th and 20th century development that, despite its unfortunate and unavoidable technicalities, it must be adequately described.

The purposes of electroplating are far greater than for ornamentation, and the two most important applications are in extracting metals from their ores and in electrolytic refining. In all plating operations the general process is the same: that if two conducting plates (electrodes) are dipped into a solution of metallic salts (electrolyte) and a current passed between them, then, under appropriate conditions of current, temperature, and so on, the metal component of the electrolyte will deposit upon the electrode carrying the outgoing current. This electrode is the cathode. The other electrode, the anode, may or may not be dissolved, depending upon its chemical relationship with the electrolyte.

In a straightforward process of ornamental silver plating, for example, the article to be plated is made the cathode, a piece of pure silver the anode, and a solution of silver cyanide the electrolyte. When a current of suitable magnitude is passed through the solution between anode and cathode, the anode dissolves, the silver concentration of the electrolyte tends to increase and, to maintain a balance, the cathode accepts a deposit of silver. It is something of this sort that forms the basis of electrolytic refining. Impure metal — let us say copper — is made the anode, and a thin sheet of pure copper is made the cathode. The electrolyte is a solution of copper sulphate, whose conditions of temperature, composition and concentration have to be carefully adjusted so that when a current is passed through the system, only copper will deposit on the cathode. The mechanism is the same as in the example of silver plating: copper is dissolved from the anode under the influence of the electric current and, to maintain equilibrium, copper is deposited from the electrolyte upon the anode. But in this case, we have the additional complication of metallic impurities in the anode, which must not be allowed to plate out. If they go into solution, then the conditions of electrolysis must be arranged so that they do not deposit on the cathode. This may involve adjusting the composition of the electrolyte, or the magnitude of the electric current. If the impurities do not go into solution, then, as copper is dissolved from around them, they will fall to the bottom of the electrolyte where they cannot cause any trouble.

So, in principle at any rate, the technique of electrolytic refining is not complicated. Where it is applicable it can produce a pure metal (which may be melted down and cast into salable shapes), as well as allowing a recovery of impurities (which may well be valuable) from the sludge or the electrolyte.

In electrolytic extraction, although the aim is quite different, the process is similar. Here the object is to produce a metal from an ore that consists of a mixture of minerals. This may be achieved by making the electrolyte of ore dissolved in a suitable solvent, and electrolyzing it so that the desired metallic constituent deposits upon the cathode. In this case the anode contributes to the process only as a means of introducing the electric current into the bath; since it must not dissolve it is often made of lead. Whereas in refining, the electrolyte acts as a sort of carrier for the metal from anode to cathode, in extraction the electrolyte is itself the primary source of metal.

But aluminum, which led us into this digression, cannot be produced electrolytically in quite this way. Although the process for which Hall and Heroult were responsible depends upon the principles that have just been outlined, it differs in that the electrolyte is not a watery one, but a molten one: it is igneous, not aqueous. A mixture of purified bauxite and other compounds is electrolyzed between carbon electrodes; the electric current both liberates the metal and keeps the bath hot and molten. This technique of igneous electrolysis produces all the world's aluminum; and a very similar process is responsible for most of the world's magnesium.

Aqueous electrolysis, however, has always had a limited application to metallic extraction — notably to copper and zinc from certain ores. But it was, and is, most successfully used in refining metals. If conditions are chosen with sufficient care, if the plant is operated with fair skill, and if there is a reasonable amount of luck (necessary in most metallurgical operations, and particularly in electrolytic ones), then only the desired metal will be plated out, while impurities may remain in solution or form a sludge.

In practice many elaborations are inevitable, and

many deviations, too, from this outline description. In a commercial undertaking there has to be provision made for circulating solutions, possibly for heating or cooling them, and frequently for treating them chemically; there have to be tanks to contain corrosive solutions; intricate electrical arrangements are necessary so that each tank can contain a maximum number of anodes and cathodes; and precautions have to be observed so that, in spite of their corrosive environment, electrical contacts will function efficiently with minimum maintenance.

The complications that so inevitably crop up in practice can make a mockery of the most ably conceived theory. For example, a story is told of the electrolytic refinery which produced good metal on the night shift, but not on the day shift. The difference defied scientific analysis, for conditions of solution temperature and concentration, electric current and voltage, and the hundred-and-one other variables that suggested themselves to the inquiring mind, were checked and found to be identical night and day. The explanation was eventually ferreted out: that the night shift foreman was a tobacco smoker, who spat into the electrolyte; and the minute alteration in the condition of the solution led to the difference in quality of the product. The story is handed down, with an appropriate dash of local color, in almost every electrolytic plant; and, although it may not be founded on fact, it is not entirely ridiculous, for additions of organic materials to the extent of less than 1 part in 1,000,000 may affect the nature of the cathode deposit. Such tiny things can affect an electrolytic process, and were learned as time went on, usually by chance.

During the latter half of the 19th century, electrolytic refining operations for copper, gold, lead, zinc and the "new" metal nickel, were introduced. This had several effects. In many cases it resulted in the ready manufacture of a purer metal than had ever been known before; and this fact itself had many repercussions, for impurities in metals can affect their properties very considerably. Ductility, resistance to corrosion, and electrical conductivity are generally reduced when a pure metal is contaminated, and although the intentional addition of foreign elements is often desirable in making alloys with specific properties (to increase the strength or hardness, for example) these are not usually the impurities that are accidentally present. It must not be forgotten that pure metals could be made before electrolytic methods were introduced, often by remelting the crude metal under a flux which removed undesirable elements. Pure copper, for example, was made in this way since the earliest times. But if the crude metal were heavily contaminated with certain impurities, this process of purification might have been difficult; and in any case, valuable constituents, such as silver or gold, were lost, for they remained alloyed with the copper. With electrolytic refining, pure copper can be made from crude metal containing any common impurities; and precious metals are easily recovered. Thus the range of useful ore deposits has been extended, for contamination with other minerals is of less significance, and the ores are more fully exploited for all the traces of other metals that they might contain.

Sometimes, however, it is more convenient to use an alternative method for refining metals; frequently it is more economical. It may, for instance, be easier to use a chemical method, by fluxing impurities; or, if the metal has a low boiling point compared with its impurities (as zinc has), it may be conveniently distilled.

The economic factors of cost and availability of power, which limit or forbid electrical practices, are obvious enough. Clearly those countries that possess water power, and have enough initiative to exploit it, are in a most favorable position; those that have a plentiful coal supply, which they use sensibly, are equally fortunate.

ZINC PRODUCTION BY ELECTROLYSIS AT TRAIL, B. C., CANADA. PILES OF RECOVERED ZINC SHEET ARE SEEN ON THE RIGHT.
Courtesy: Consolidated Mining and Smelting Company

Electrolytic refining and extraction, when they were introduced, were quite new techniques, forming a divergence from the general line of metallurgical progress. The other exploitation of electricity in the metallurgical industry was in electrical heating, which, although a new weapon, in some respects continued the old tradition; it was rather in the nature of a new handle to an old broom. Nevertheless, electrical heating has useful applications, which are valuable, but not vital; and are limited economically. Among the most important are its uses in steel manufacture, both as an actual process of steelmaking and in producing ferroalloys. Some metals have such high melting points that electrical methods of melting are the only possible practical method. For instance, titanium is almost exclusively melted by the electric arc. There is an alternative method that involves a high frequency electrical current. However, the technique is generally ignored, misunderstood or maligned.

But even to outline all that electrical methods of heating involve would add still more technicalities to a chapter that is already sufficiently textbook.

While they are still in their infancy, it would be imprudent to forecast the future of electrical methods. They are already invaluable in refining, vital in certain extractions, and desirable in some melting processes; and it seems probable that they will be used more as time goes on. But to do this with any success will demand a careful survey of fuel resources, thorough methods of using the heat contained in the belching smoke of factory chimneys, and a sensible exploitation of water power.

CHAPTER 11

1855–1957 A.D.: Iron and Steel

MORE SPECTACULAR than the steadily increasing application of electricity to metallurgy was the dramatic introduction in the late 19th century of two large-scale steelmaking processes, which have always been associated with the names of Bessemer and Siemens. If we accept American claims we should pay an equal tribute to William Kelly; and if we wish to be precise as well as just, we should add the names of several other people who contributed to these new methods of steelmaking.

Steel production depends entirely upon a ready supply of iron from the blast furnace, and it was fortunate from this point of view that blast furnace practice had advanced sufficiently to meet the new demands. The capacity of the iron blast furnace had been increased enormously since its early days, and it had been given a form similar to modern design. A furnace that was erected in 1861 at the Congreaves Iron Works in Staffordshire can be taken as representative of an up-to-date construction for those times. The internal shape of the furnace was rather like an elongated football with its ends cut off; it was about 50 ft. high, built of brick and braced with iron hoops; and it had an internal diameter at the widest point of its belly of 17 ft.; the hearth diameter was 10 ft. Four wrought iron tuyeres with hollow casings, through which cooling water was circulated, entered at hearth level. The choice of refractory bricks for the furnace was not judicious according to modern views, but it did at least show that ironmakers were becoming conscious of a problem. The approximate output of such a furnace was about 50 tons of iron each week.

When the molten iron was tapped from a furnace, it was directed into channels made in sand, and ran along these into smaller furrows that branched off from the main stream. From the similarity of this arrangement to a sow feeding her litter, the iron that solidified within the main channels was known as "sow," and that in the subsidiary ones as "pig." These names are still standard in the colorful vocabulary of the ironmaker. After the iron has solidified the pigs are broken off and the sows cut up into smaller pieces. The material is known as "pig iron." Quite clearly, if such iron is remelted and cast (for example, into the shape of a fire grate), it still remains from a point of view of composition pig iron, unless it has been refined in any way. But it is usual to call iron that has been purposely cast into a shape for some specific object "cast iron." The distinction between cast iron and pig iron need not involve any difference of composition, but only a difference of purpose.

Pig iron from the blast furnace has a composition that naturally depends very largely upon the composition of the ores used in its manufacture. It generally contains 3.00–4.0% carbon dissolved from the coke, and accounting for the low melting point of pig iron. Then, depending upon the composition of the ore, fluxes, and the way in which the furnace is run, it would contain:

0.02–0.2% sulphur	0.40–5.0% silicon
0.02–2.0% manganese	0.02–2.0% phosphorus

Many modifications were made to blast furnace design and operation, but probably the greatest single innovation was the introduction of the hot air blast. Although it seems very natural that somebody should have realized that a cold blast would contribute largely to the inefficiency of a furnace, and that more heat might be saved in the furnace than would be used in pre-heating the air, these were not the motives that prompted the introduction of the hot air blast.

It was quite a popular belief among ironworkers that the erratic behavior of furnaces during the summer months could be ascribed to the presence of sulphur in the air. James Neilson, to whom this was suggested, preferred to believe that the true cause was not sulphur, but a lack of oxygen and an increase in humidity in the summer air. He therefore suggested that the air should be dried by passing it through iron tubes containing quicklime.

But, before he put this into effect, he was fortunately confronted with another problem that led him along different lines of thought. At a particular ironworks it was found that a furnace placed

A MODERN BLAST FURNACE INSTALLATION.

some way from the blowing engine did not work as efficiently as other furnaces closer to them, and Neilson ascribed this to a poor air supply. Heating the air, he suggested, would increase its volume, and might put the matter to rights. In experimenting with the idea, he found that, whatever else it did not do, the heated air intensified combustion; and he suggested that preheating should be applied to the furnace blast. This conflicted with the popular opinion that the best iron was made with as cold a blast as possible — after all, was not good iron always made in the winter? However, the invention, patented in 1838, was eventually accepted, and put into effect by passing the air through cast iron pipes that were heated by coal or coke, and into the furnace by water-cooled tuyeres. Before many years had passed, the hot waste gases from the top of the blast furnace were used to heat the iron preheater tubes; and this contributed to the efficiency of the system. By using an air blast preheated to 800°F. (a temperature dictated by the deterioration of the cast iron pipes above this figure), and by other more incidental efficiencies, the weight of coke to produce a ton of pig iron was reduced approximately from 6 tons to 2 tons.

Neilson's idea of *drying* the air for the blast furnace has not gone unnoticed, either. In 1904, James Gayley constructed a refrigeration unit to take the moisture from the air. A 1000-ton per day furnace takes in as much as 60 tons of water each day; but the reason for wishing to exclude this is not so much that the water increases coke consumption (for opinions differ on this contention), but that variations in the air's humidity lead to variations in the behavior of the furnace. "The blast furnace operator," said John L. Bray, "watches the weather with an ever keener interest than the baseball club manager."

Although Gayley established that the dry blast eliminated operation variations, his refrigeration system was not economically sound. However, more recent developments involve the removal of water

A. OPEN HEARTH FURNACE
A. I ROOF
A-2 HEARTH
A-3 BATH
A-4 BURNER
A-5 DOOR
A-6 CHECKER CHAMBER
B. LADLE ADDITION CHUTE
C. SPOUT
D. STEEL LADLE
E. SLAG THIMBLE
F. INGOT MOLD
G. INGOT BUGGY
H. POURING PLATFORM
I. STOPPER ROD ASSEMBLY
J. POURING LADLE CRANE
J-1 MAIN HOIST
J-2 AUXILIARY HOIST
K. SPOUT CRANE
L. HOT METAL CRANE
L-1 MAIN HOIST
L-2 AUXILIARY HOIST
M. HOT METAL LADLE
N. CHARGING MACHINE
O. CHARGING FLOOR
P. CHARGING BOX
Q. CONTROL PANEL
R. CRANE CAB
S. ELECTRIC VALVE HOIST
T. DUCT
U. WASTE HEAT BOILER
V. STACK DAMPER
W. STACK
X. ELECTRIC DOOR HOIST
Y. REPAIR HOIST
Z. SLAG POCKET

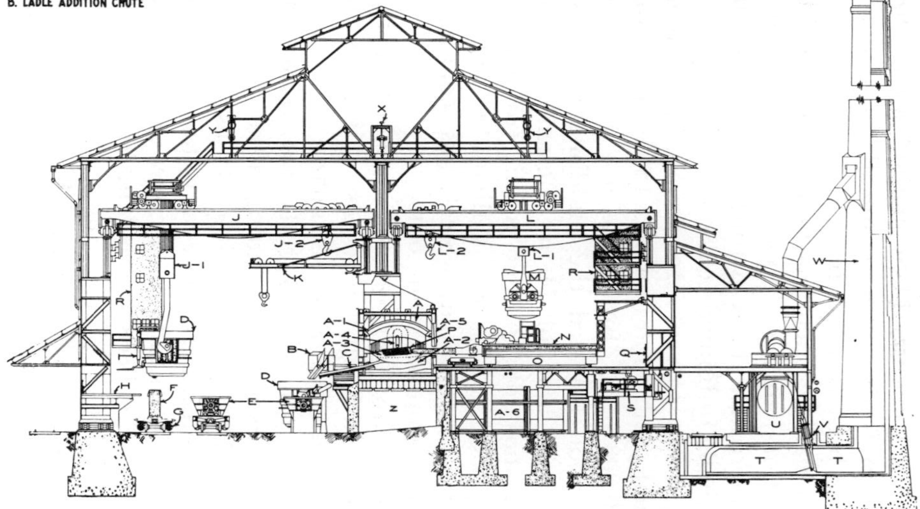

CROSS-SECTION THROUGH AN OPEN-HEARTH PLANT.

vapor by condensation in a chilled water system, or by chemical dehydrating agents to give a constant moisture content. These methods appear to be both metallurgically and economically good practice.

The blast furnace, by the mid 19th century, was a reasonably efficient machine. Stimulated by Cort's inventions, thriving upon the demands of industry for more and more wrought iron, and fed by a seemingly unlimited coal reserve, European pig iron production soared. In Britain, for instance, the annual production in about 1860 was approximately 4,000,000 tons. But it was not for long to remain the spoon that fed the puddling furnaces, for by 1860 the fame of Henry Bessemer's process had spread not only throughout the country, but across the world; and by 1870 William Siemens' name had become as familiar. Steel was coming in, and wrought iron was going out.

Henry Bessemer was not a steelmaker, but a most remarkable English inventor. His early inventions were connected with an amazing diversity of subjects: a machine to emboss Utrecht velvet; improvements to type founding; a stamp for documents that could not be fraudulently used; the manufacture of bronze powder, varnishes and paints; improvements to railway carriages (one presumes they have not yet been effected in Britain); the ventilation of mines; pumps; sugar manufacture; optical glass making; projectiles. The list seems endless. Bessemer's patents alone — and some of his inventions he did not protect in this way — numbered over 100 between the years 1838 and 1883; and his fruitful ingenuity is perhaps accounted for by his own words: "I find that persons wholly unconnected with any particular business have their minds so free and untrammelled to view things as they are, and as they would present themselves to an independent observer, that they are the men who eventually produce the greatest changes." No doubt Bessemer had himself in mind.

It was a direct consequence of his work on guns and projectiles that Bessemer set out to improve the quality of cast iron, and in 1854 he applied for the patent for "Improvements in the Manufacture of Iron and Steel" that was to be the forerunner of so many others that represented both disappointments and triumphs. The process that Bessemer suggested was to melt pig iron in a reverberatory furnace, and to decarburize it by blowing air over the surface of the metal. Had Bessemer continued along these lines he might have been responsible for the other great steelmaking technique — the open-hearth process. But his attention was diverted by the fact that a hot bar of iron exposed to an air blast became quite decarburized, and to exploit this effect Bessemer made his first "converter." This was an egg-shaped brick vessel charged with 10 lbs. of molten pig iron and heated by a coal fire. A fireclay pipe was introduced through the top of the converter into the surface of the metal, and, after air had been blown through it for half an hour, the iron had become quite free of carbon; the other impurities had become oxidized away; and an ingot of fairly pure iron was cast. Encouraged by these results, Bessemer set up a second and larger converter in the garden of his premises at St. Pancras, London,

EARLY AMERICAN BESSEMER CONVERTERS. THE CONVERTER AT THE RIGHT IS BLOWING, WHILE THE OTHER IS IN THE CHARGING POSITION.

Courtesy: American Iron and Steel Institute

chiefly differing in that there was no source of external heat, for Bessemer believed that the heat generated by the oxidation process that went on within the iron might be sufficient to keep the iron molten. In this he was not mistaken. The air was now introduced through tuyeres at the bottom of the vessel, and molten pig iron was charged through a trough at the side. Bessemer confessed that the results were a revelation, and the sheet of flame and showers of sparks that leaped from the mouth of the converter so disconcerted him that in his next experiment he hung a cast iron plate a few feet above the top of the converter, to lessen the danger of fire caused by the eruption of hot cinders. The plate, far from fulfilling its purpose, melted in a few minutes, and with it any fears that Bessemer entertained on the process not being self-heating.

The first public announcement of this great discovery was made in August 1856, when Bessemer read a paper on "The Manufacture of Iron without Fuel" to the British Association. The meeting did not at first take the paper seriously, but before Bessemer had finished, it was convinced of the tremendous importance of the process. That Bessemer "sold" his process is indicated by the fact that, shortly afterwards, five independent firms bought licenses from Bessemer—and paid him a total of £27,000. He showed that the process gave an iron that could be cast into large ingots; and, most remarkable of all, that if the air blast were checked before all the carbon had been oxidized, the process yielded what he called "semi-steel." Although the vast potentialities of this last claim were not immediately appreciated (and, in fact, were not so easily realized in practice as Bessemer hoped), the invention created nationwide excitement. Even *The Times* (London) published a full account of Bessemer's paper.

But these high hopes and enthusiastic beginnings were not long-lived, for trials at ironworks in many parts of the country were disastrously unsuccessful, and they yielded a brittle, unserviceable iron. Acclamation soon turned to derision in a not unprecedented way, and the recently heralded process was described as "a brilliant meteor that had flitted across the metallurgical horizon for a short space, only to die out in a train of sparks, and then vanish into total darkness."

The story of Bessemer's obstinate struggle to find and correct the cause of the failure is one of dogged persistence in the face of criticism and ridicule – a record one tends to forget in summarizing his conclusion. The trouble was due to the high phosphorus content of the pig iron that was used. Although most of the impurities of pig iron were eliminated by oxidation, phosphorus was unaffected; and it was only fortunate (or unfortunate, perhaps) that the materials used in Bessemer's first trials were phosphorus-free. The manufacturers who met with failure used iron made from British ores which had a high phosphorus content that was removed neither in the blast furnace nor in the converter; and this made Bessemer iron brittle and useless. The puddling process, on the other hand, effectively removed phosphorus, and puddled iron was satisfactory no matter what kind of pig iron was used in its manufacture. Discovering this explanation by a systematic chemical analysis, Bessemer imported Swedish pig iron, free from phosphorus; and the process was again put into operation in 1858. But the ironmakers and steelmakers were too disbelieving to accept it.

Characteristically, Bessemer established his own factory in which he produced, not malleable iron, but steel; for he had realized the greater potentialities that this material possessed, if it could be cheaply made. Characteristically too, he set up his works in the heart of Sheffield industry, which still stubbornly adhered to the Huntsman process; and soon he made steel at £15 a ton below the price of his competitors.

But this is not the end of a remarkable story. Bessemer searched Britain for suitable ores — and found them. In 1858 he improved his converter by mounting it on pivots so that it could be tilted to charge and empty it. Although good steel could now be made from British ores, there was still an enormous prejudice against the process. This is to some extent explained by its initial spectacular failure, but more fundamentally by an unfamiliarity with the proper use of steel.

Before 1858, steel was so expensive as to be very restricted in its application; but the Bessemer process reduced its price sufficiently to allow its use in large quantities — as plates, rails, girders and so on. Workmen were not accustomed, however, to the new material or to the new techniques that it demanded, for steel, essentially an alloy of iron containing anywhere from 0.1 to 1.5% of carbon, undergoes internal structural changes when it is heated and cooled. Therefore any fabrication process that entails hot working has to be carefully analyzed and skillfully controlled. This was not appreciated at the time. Although metallurgists knew that steel seemed more temperamental than iron in its behavior, they did not altogether know why; and steelworkers doggedly employed the techniques that they had learned in using iron. As a result, steel was badly misused, and it was common for its misusers to ascribe failure to the fictitious shortcomings of the Bessemer process, rather than admit their own incapabilities.

By 1875, the annual production of Bessemer steel

in England had reached over 700,000 tons, and many steel users, railway companies and boiler makers put the metal into service. Nevertheless, and in spite of the popularity of the process abroad, the British government would not accept Bessemer steel in any of its contracts. Despite the satisfactory behavior of the steel under test conditions, they stubbornly refused to use it for guns and shipping, while other countries exploited the English invention.

The fact that the metal was satisfactory for ship construction is shown by the fact that private and company-owned ships in England were made of Bessemer steel from 1863 (when the 329-ton steamship *Pelican* was launched), and in succeeding years the names of increasing numbers of bigger steel ships appeared on *Lloyd's Register.* Nevertheless, as late as 1875 the Chief Naval Architect of the Royal Navy, in a paper on "Iron and Steel for Shipbuilding," alleged "uncertainties and treacheries of Bessemer steel in the form of ship and boiler plate," and he recommended the continued use of iron. Iron ships had become quite common since 1860, and although they had obvious advantages over the previous wooden or copper-sheathed vessels, steel, on account of its greater strength, enabled lighter ships to be built and allowed a consequent increase in freight. Bessemer stormed over the provoking "uncertainties and treacheries." But his anger blew itself out, for, by 1885, Bessemer steel was accepted by the British government both for guns and in naval shipbuilding.

About one year after Bessemer's announcement of his process, Göran Frederick Göransson bought the rights to manufacture Bessemer steel in Sweden.

After unsuccessful attempts with a converter that Bessemer had supplied, Göransson made his own converter, and on July 19, 1858, made good steel — by opening the tuyeres and so considerably reducing the air pressure below the then accepted figure and increasing the volume of air introduced to the molten bath.

Göransson was able to check the carbon content of his melt by a rapid test introduced by Professor Eggertz, and hence was in a position to stop the blow at the precise moment at which steel of the correct carbon content had been made. In the absence of a quick carbon check, stopping the blow at the right time offered a most exasperating problem.

At the time of Göransson's success, things were not going well with Bessemer, who was trying to make his process reliable by granulating the product, sorting the steel into different carbon grades, and remelting it. This, clearly, was not economically pleasing. Göransson's efforts did, at least, show that the Bessemer process had great potentialities and could not be written off as a dying train of sparks. However, Göransson did not have the advantage of raw materials that were not only low in phosphorus, but high in manganese. For even when Bessemer used low phosphorus pig, he produced a steel that was hot-short. It was brittle when it was hot and so could not be forged.

In 1856–57, when Bessemer was at the depth of his troubles, Robert Mushet was working on the effects of spiegeleisen (a pig iron containing 15–30% managanese and 4–7% carbon) in reducing hot-short tendencies. A complete account of his method was filed at the British Patent Office in March 1857. Applied to Bessemer's process, Mushet's technique would involve blowing the iron to completion and recarburizing with spiegel — which would not only bring the carbon to the required level, but would introduce manganese and so prevent hot-shortness.

For rather obscure reasons, Mushet did not divulge his remedy to Bessemer who, it would seem, found it out for himself by reading of Heath's work on a similar subject. (And even before Heath, William Reynolds in 1799 filed a patent for the use of spiegel.) Mushet's reticence resulted in a substantial financial loss; but Bessemer later paid him an annuity of £300. It is sometimes suggested that this implied an obligation to Mushet; but if that were the case, it is hardly likely that Bessemer would have been the enthusiastic supporter he was of the award to Mushet, in 1876, of the Bessemer Medal — unless this, too, is an indication of a guilty conscience.

In any case, spiegel had its limitations, for its low ratio of manganese to carbon made it ineffective in producing low-carbon steels — insufficient manganese being introduced. The production of ferro-manganese (a material with a higher manganese to carbon ratio) was largely due to Bessemer's efforts.

These troubles over, the Bessemer process was fairly firmly established — a process that would yield 5 tons of steel from molten pig iron by simply blowing air through it for 20 minutes; that reduced the price of a ton of cast steel from about £50 in 1850 to £10 in 1860; that in 1882 had been adopted by over 100 works throughout the world, employing 3300 converters producing more than 4,000,000 tons

ROBERT MUSHET, WHO CONTRIBUTED TO THE SUCCESS OF THE BESSEMER PROCESS, SHOWN HERE WITH HIS WIFE.

Courtesy: Steelways

of steel. Henry Bessemer was understandably self-satisfied when, in 1861, he made the prediction that "as certain as the age of iron superseded that of bronze, so will the age of steel succeed that of iron."

Before such vast increases in production had been achieved, it was necessary for the outstanding limitation of the Bessemer process to be overcome — that steel and malleable iron could be made only from nonphosphoric pig iron, which, in turn, demanded nonphosphoric iron ores. Although there were fairly extensive ore deposits throughout the world that would produce suitable iron, phosphoric ores were much more common. Before a technique had been evolved that would allow their use, the open-hearth steelmaking process had been introduced by Siemens, and this suffered from the same deficiency. The eventual solution to the problem resulted, therefore, not only in greater production figures for Bessemer steel, but in the expansion of the open hearth process, which, before long, became more popular.

Charles William Siemens, who was born in Hanover, Germany, in 1823, like Bessemer was an inventor — but one whose inventions depended more upon a development of a scientific idea, than upon the possible exploitation of a commercial necessity. He settled in England in 1844, and after working on electrical processes, printing, pumps and other things, he turned his attention to reverberatory furnace heating.

By this time it had become common for the reverberatory furnace to be gas heated: burners at one end of the furnace directed flames across the hearth so that the charge was heated both directly and by reflection from the furnace roof. The burned gases went up the chimney stack at the end of the furnace opposite the burners, where their considerable heat content was allowed to go to waste. In an endeavor to produce higher temperatures in the furnace, Siemens, at the suggestion of his brother, utilized this heat content to preheat the incoming air and gas. The burned gases were passed through two brickwork chambers (known as regenerators) loosely packed with refractory bricks. The bricks became hot, and after half an hour or so the direction of gas flow across the hearth was reversed: gas and air were admitted separately through the two hot chambers on their way to the burners, and left the furnace via two further regenerators which, at the next reversal, became the preheaters. In this way some of the heat from the process was returned to the system, and this efficient practice applied to gaseous combustion produced a higher temperature than had previously been attainable in the reverberatory furnace. Therefore, its first use was suggested in glass manufacture; and when Siemens patented his process in 1861, he foresaw its application to steelmaking. (Within a year or two Cowper had applied the regenerative system to ironmaking, alternately passing air for the blast and the waste gases through brickwork chambers, and effected a further saving of coke.)

The regenerative furnace was in operation for glassmaking in 1861, and during the following years a number of ironmasters tried to make steel in it, but without success. In 1867 Siemens erected an experimental plant of his own, and before long he was able to make good steel from a charge of scrap iron and iron ore. Concurrently, Emil Martin in France, who used the furnace under license from Siemens, made steel from a charge of wrought iron, steel scrap and pig iron. The essence of both techniques has remained unchanged. At the high temperatures reached the charge melts and the carbon and other impurities are removed by the oxidizing influence of the iron ore and fluxes that are added. When the iron is almost decarburized, additions are made of spiegel and other materials to produce a steel of appropriate composition.

In 1878 Emil Martin and his brother Peter claimed the success of the process (the open-hearth steelmaking process) as their own, an assertion that led to a protracted battle finally arbitrated by an impartial body of metallurgists. The gist of its findings was that the process was Siemens', and he alone was to benefit from royalties, licenses and so on. However, many improvements were due to the Martin brothers, and their value can be judged by the fact that the method of steelmaking is frequently called the "Siemens-Martin" process.

There has been no such agreement, however, on the part played by William Kelly of Kentucky in the manufacture of steel by the Bessemer process. Kelly's champions maintain that he worked at a "pneumatic process," identical to Bessemer's, from 1847. The process was not altogether satisfactory; but with Mushet's patent in the United States for the use of spiegel, in 1857, Kelly was able to produce good quality steel. He applied for a patent only a few days after Bessemer; and as the U.S. Commissioner of Patents was convinced that Kelly had evolved the scheme some years previously, the patent was granted.

Kelly's venture was eventually merged (after yet another confused legal battle) with that of Alexander Lyman Holley's. Holley had obtained a license from Bessemer in 1862 during a visit to England; he returned to the United States to install his plant.

Commercial production of converter steel (perhaps in this context we may call it Kelly steel) began in the United States about 1866.

Nevertheless, Bessemer achieved the honor and the glory — and most of the royalties; and one cannot help agreeing a little with Kelly's supporters that this was due rather to Sir Henry's hard-headed business methods, than to his genius being rewarded.

Perhaps we cannot concur so enthusiastically over another claim. Kelly, so the story goes, employed two Englishmen to help him with his work when it was still in its experimental stages; and despite their apparent honesty, they disappeared quite without notice after gaining a thorough working knowledge of the principles of the "pneumatic process." Although of an unsuspecting nature, it is not surprising that Kelly was doubtful of their intentions, and as a result of investigations he was fairly certain that they had returned to England. Any misgivings he may have entertained were, it is said, more than confirmed when, some years later, he saw a photograph of Bessemer and recognized it as an exact likeness of one of the two absconding workmen. With this belief it is not surprising that Boucher concludes his book, *William Kelly: a True History of the So-Called Bessemer Process,* by preaching: Let his [William Kelly's] accomplishments impel the sculptoring hand of the future to erase the name of Henry Bessemer from the cold and lifeless marble of the past, and carve in its stead the name of William Kelly.

Siemens or Martin, Bessemer or Kelly (or even Popov, perhaps) — no one had so far found a way of making steel from phosphoric pig iron; for the open-hearth process, like the Bessemer process, did not effect any removal of phosphorus. The reactionary supporters of wrought iron as a constructional material held a trump card, for the world's iron ore deposits are predominantly phosphoric, and to what other purpose could these be put than to make wrought iron? This was the alarming problem that beset the metallurgists of the 1870's, for realizing that steel had come to stay (not that metallurgists are always so progressive), a method of using phosphoric ores had to be found.

One of many papers on the subject of phosphorus removal was presented in September 1878, by Low-

OPERATING AN EARLY OPEN-HEARTH FURNACE (ABOUT 1870).
Courtesy: American Iron and Steel Institute

thian Bell, at the Iron and Steel Institute's rooms in London. During the discussion that followed, when it was generally agreed that phosphorus could *not* be removed on a commercial scale, Sidney Thomas rose to say that he had, in fact, found a way to do this. But Thomas was a young man, and his observation was quite ignored.

Thomas, who was by profession clerk at the Thames Police Court, and his cousin Gilchrist prepared a paper on "The Elimination of Phosphorus," which was discussed in May 1879, by which time the authors had put their proposals into successful operation at a Middlesborough steelworks. Thomas' interest in the subject began only in 1870, when he deduced that phosphorus was not removed in the Bessemer converter because of the siliceous nature of the converter lining and of the fluxes added. With Gilchrist, a chemist, he studied and experimented, and by 1878 had shown that by using a lime flux and a converter lining of magnesian limestone (both of these are basic materials, in contrast to the acidic nature of the previously used lining and flux) the removal of phosphorus and the formation of phosphates was encouraged. A metal free from phosphorus could be produced by blowing the metal in the converter for a few minutes after the main reaction had subsided. The slag that resulted was so high in phosphates as to be a valuable fertilizer. This, the Basic Bessemer process, is still used by works that make steel in the converter from phosphoric pig irons.

The discovery of the basic method was very soon applied to the open-hearth process; and the 1880 picture is completed by the small bright streak of electric furnace steel making, due to Siemens, and the fading colors of the crucible process.

Steel production increased, with its temporary ups and downs, from 1,750,000 tons in 1875 to a present (1957) annual production of more than 200,000,000 tons. Production curves over the years tell their story of booms and slumps, of countries losing and gaining places in the steelmaking race. Britain inevitably lost her lead in 1890, and the United States, at the front since that time, produced in 1955 more than 100,000,000 tons of steel representing about one half of the world's total output. The Bessemer process, responsible for almost all the steel made in 1875, became secondary to the open-hearth process before the end of the 19th century, and today little more than 10% of the world's steel is made in the converter; 2 or 3% is made in the electric furnace; and only a very little is made in the crucible.

An approximate analysis of figures for 1955 indicates that in western Europe about 40% of the steel is made by the Bessemer process. Since European ores are predominantly phophoric, it is not surprising that 98% of western Europe's Bessemer steel is made in the basic converter. In the United States about 2.5% of the total steel production is made in the converter; almost all is made by the acid process.

Electric furnace steel is usually made by the arc method – in which the charge (usually high in scrap steel) is melted by the heat of a carbon arc and subsequently refined. The process, due to William Siemens, was introduced in the late 19th century and developed by Stassano in Italy, Heroult in France, and Hall and Acheson in the United States. Since those days it has shown a continued expansion. In the United States, where the percentage of electric steel manufactured is highest, it represents about 10% of the total production. Another steel melting method using the electric current depends upon the heat generated in a body of metal surrounded by an alternating electric field. This technique involves costly equipment and is used only in the production of high quality alloy steels.

The present day attitude towards converter practice is interesting. Until five or ten years ago the process was looked upon disfavorably outside Europe. Sometimes the claim that Bessemer stock was of poor quality could be substantiated; often the opinions were prejudiced. Additionally, much American pig iron contains an amount of phosphorus that is too great for acid practice, but is insufficient to maintain an after-blow.

However, there is no doubt that Bessemer steel producers became increasingly aware of two deficiencies of the process: first, the inevitable nitrogen pick-up in the steel from the air blast; second, in basic practice, the difficulty of judging the end of the after-blow. Although experienced operators claim that they can judge the end of the after-blow by the appearance of the flame and of a sample taken from the furnace, there has always been considerable interest shown by European manufacturers towards more objective assessments. Color analyses of the bath and many different methods of analyzing the flame characteristics are still being developed with a view to providing better control. In a method due to Perrin, the after-blow is eliminated completely by pouring the steel (still containing phosphorus) into a highly aluminous slag. This process — Perrinization — certainly eliminates the uncertainties of the after-blow, but adds expense.

The elimination of the nitrogen content in Bessemer steel has been attempted in three broad ways. The first method, introduced in Britain by Dickie, involved a careful control of the blow to reduce the amount and partial pressure of nitrogen in contact with the steel. This was facilitated by iron ore additions that introduced oxygen to aid the refining process, but did not of course supply nitrogen. A second method, used in Germany, involved tilting the converter so that the air travels a shorter distance through the bath. This principle is also followed in the Tropenas converter (which is described later). The third method seems to have the most potentialities: the blow is made by an oxygen-enriched blast or by a blast of pure oxygen. This was, in fact, anticipated by Bessemer himself, for in the contemporary Linz-Donawitz process, developed in Austria, oxygen is introduced into the bath by a lance inserted from the top of the vessel in a way that Bessemer described in a patent filed in 1855. At present the Linz-Donawitz process is applied only to low-phophorus pig since a bulky slag interferes with the oxygen blast and, in any case, the end of the reaction is difficult to detect when phophorus has to be removed.

This survey of steelmaking in converters will be concluded by an outline of the work of Alexandre Tropenas, a Frenchman, who worked on an in-

THE FIRST AMERICAN CONTINUOUS HOT-STRIP MILL, AT ASHLAND, KENTUCKY, 1923.

Courtesy: American Iron and Steel Institute

vention made by Robert in 1886. Robert — a Frenchman, too — investigated the use of an air blast set tangentially to the surface of the molten steel. (Bessemer, by the way, appears to have foreseen this, too.) Among many advantages claimed for the Tropenas converter, perhaps the principal one is that it is easily adapted to a small-scale operation. While the contemporary Bessemer type of converter has a capacity about 25 tons, the Tropenas may hold as little as half a ton. Consequently it is suitable for foundry use. The side-blowing method used in the Tropenas vessel also appears to produce a steel of lower nitrogen content than conventional bottom-blowing.

Turning our attention to recent developments in ironmaking, these include drying, or partially drying, the blast — as we mentioned earlier; and the possibilities of enriching the blast with oxygen are under active investigation. In 1910 Jernkontoret, in Sweden, built the first electric furnace for the manufacture of pig iron. Power consumption in the furnace is very high (2000–2400 kwhr per ton of pig produced). Consequently more interest has been shown in the electric method that produces not a molten pig iron, but a sponge iron. This process, too, was developed in Sweden, by Sieurin, about 1910, and involves reducing ore in a shaft furnace by a mixture of carbon monoxide and hydrogen introduced at the bottom of the furnace. The upper part of the furnace burden is heated by the combustion of these gases; the lower part is heated electrically, and contains coke whose purpose is to convert any carbon dioxide and water vapor (in the introduced gases) to carbon monoxide and hydrogen. The ore charged to the furnace must be rich, since there is no slag separation. The iron content of the sponge is about 90%; the remaining 10% is mainly oxygen. The present interest shown in the process is largely due to the fact that it requires only one third of the normal blast furnace coke consumption, and the electric power required is not too high — about 1000 kwhr per ton.

Any summary of the present day position of iron and steelmaking is likely to provoke a storm, and perhaps the briefest conclusions will be the most acceptable. First, ironmaking techniques are under active review; but it is difficult to believe that the conventional blast furnace will be widely superseded, even though it may be modified. As to steelmaking, it is fairly safe to assert that the crucible process, although it is still occasionally claimed to produce steel of unsurpassed quality, will never be revived. It contributes to the world's output an ever decreasing drop to an ever growing ladle. Electric furnace steel, nearest in quality to crucible steel, is used when occasion demands it, or where power resources allow it. It is particularly suitable where the raw materials are constituted of scrap. But in the absence of any wild discoveries, converter processes and open-hearth processes are likely to provide the bulk of our steel for some time to come. The decline of the converter processes may well be checked by the introduction of refined techniques – particularly the oxygen lance method.

Just as the uses of the alloy, bronze, overshadowed the utility of its predecessor and major component, copper, so the advent of steel has had much more striking effects than its parent iron. In more ways than this, the introduction of steel in our age is a parallel to the discovery of bronze in an earlier one; and today we are experiencing changes in our habits and lives, as earlier civilizations did when they became familiar with the manufacture of bronze. To us, of course, the introduction of steel seems to have far greater effects than we imagine the discovery of bronze to have had – a child is far more impressed by the rock pool he plays in than in the receding sea that produced it. We may look back and analyze the results of bronze culture, for its contribution to the growing structure of civilization is firmly placed and thus readily interpreted: but we can only superficially appreciate the effects of a steel age, for we are living in the midst of the effects we are producing. So, since we are history makers as regards this period, final evaluation of our contribution will have to be made by future historians.

CHAPTER 12

Alloy Steels and Superalloys

AT THIS POINT in the history of metals it would be futile to try to develop the story further along chronological lines. The last 50 years — particularly the last 20 years — have brought with them so many changes in production techniques and engineering applications of metals that any chronological account would be a day-to-day one. Of course, some of the more recent innovations assume an exaggerated importance because of their proximity; but we, after all, are feeling their exaggerated effects.

Since the introduction of large scale steelmaking 100 years ago, users have demanded that steels are available to withstand increasingly exacting conditions. An engineer no longer merely states that a component is to be made of steel. He specifies, or should specify, the exact type of steel to be used. And while most steel sections used in building construction are made of simple plain-carbon alloys (much the same sort of material as was made 100 years ago), even these are clearly specified, and tolerances of "impurities" laid down so that exact mechanical properties can be forecast.

However, conditions and economy often demand alloy steels with superior qualities. The conditions demand them if the steel component is, for example, to be subjected to severe loading or to a corrosive environment. Economy may demand alloy steels, when a smaller section leads to lower tonnages and hence to reduced freight rates and labor costs. Among the greatest contributors to the usage of alloy steels are automotive, aircraft, chemical and power-producing industries. An average car contains nearly 300 lbs. of alloy steels — in transmission, gears, springs, steering mechanism. In aircraft the problem is to minimize weight without sacrificing strength, which demands the use of small, strong components. The chemical industry needs alloy steels for use in parts ranging from oil well drilling equipment to plastic extrusion machines. And in power production the turbines, high-pressure pipes, drive shafts, and so on would be impossible to make, or so cumbersome as to be impracticable, if alloy steels were unavailable. Many of the machine tools of these industries are made of alloy steels. At present, of the United States steel production, about 10,000,000 tons annually (representing approximately 10% of the total production) is in alloy steels. In Britain, about 1,000,000 tons (representing 5% total production) is in alloy steels.

One of the principal purposes of alloying elements is to increase the strength of the steel. Nickel, chromium, molybdenum, vanadium, tungsten (or, as we are supposed to call it, wolfram) are among the most common additions. To withstand a given load, a plain carbon steel of very substantial dimensions might be theoretically possible; but it would be inconvenient. And, of course, the plain carbon steel could be strengthened to some extent by cold working; or it may be heat treated to improve its properties. But heat treatments are restricted since, on quenching, a plain carbon steel does not harden to a very great depth; the outside is hardened, but the center is unaffected by quenching. For example, a 0.6% carbon steel 1-in. round bar will not harden at its center if it is water quenched. With appropriate alloy additions complete hardening can be obtained; and, more than that, deep hardening will result even when slower cooling rates are employed. Thus, a large component which would be difficult to quench, say, in water and which, in any case, would not harden throughout, may be suitably alloyed so that it will harden on air cooling.

Not only strength, but toughness (that is, resistance to failure by impact) and wear resistance are improved by the correct alloy additions. Plain carbon steels are particularly prone to failure at subzero temperatures. Alloy steels have been devised that possess far superior properties.

When metals corrode, they usually cause the most phenomenal financial setbacks. Not only must the cost of the metal be considered, but also the cost of shutdowns due to any damage from the corrosion failure, and of labor costs involved in replacement. Corrosion losses cost the United States billions of dollars for replacement alone; and "indirect" losses swell the figure further. While it may be argued that this sort of loss and replacement maintains a healthy economy, it is difficult to convince a steel user of the validity of such a claim; and the metallurgist is much more inclined to recommend the use of materials that will not corrode. The corrosion

SMALL PRESSURE JET ENGINES MOUNTED TO THE TIPS OF HELICOPTER ROTOR BLADES ARE ENCASED IN HASTELLOY X — A "SUPERALLOY."

Courtesy: Haynes Stellite Company

resistance of steel can be improved to some extent by addition of about 0.25% copper, alone or with 0.2% phosphorus. More severe conditions may demand the use of nickel, chromium and molybdenum in some combination and to a total figure of approximately 10%. The stainless steels contain even larger amounts of chromium, often with other elements. The most common types of stainless steel may contain 12–30% chromium, or 6–40% nickel with 16–30% chromium – and, possibly, 2% molybdenum and other odds and ends.

With the development of the jet engine, alloys were demanded that would withstand the severe conditions of the corrosive environment, high temperature and high stresses that pertain at the blade of the turbine. Even stainless steels were not adequate; and the range of materials generally known as "superalloys" was developed. The blades of a turbine, it will be appreciated, are subjected not only to high-temperature corrosion, but to a continual stress (due to centrifugal force) which, at the high temperatures, cause the metal to "creep." It stretches plastically, rather like a piece of chewing gum. Alloys composed of a distressing complexity of elements have been evolved that will withstand these conditions. They are usually based on a composition high in nickel and chromium, or high in cobalt, nickel and chromium; but the small additions of other elements (aluminum, titanium, carbon, molybdenum, tungsten) are as important as the pinch of salt on a potato chip.

Now, these developments have demanded a production of "new" metals – metals that have not so far been considered in this book, mainly because their appearance on the metallurgical scene is very recent. But it would be tragic if the impression had been given that metals like nickel, chromium, molybdenum, vanadium, tungsten, cobalt were unimportant simply because they are precociously young. It is a metallurgical nightmare to try to imagine, for instance, a world without nickel.

The introduction of alloying elements into steel is usually (but not always) accomplished by the use of "ferro-alloys." Nickel is usually added "straight"; but manganese, chromium, vanadium, molybdenum, tungsten (and silicon for purposes of deoxidation) are thrown into the furnace, or ladle, in the form of an alloy of the appropriate element with iron. Ferro-alloys are generally made in a small blast furnace or in an electric arc furnace.

Apart from their uses in steel and superalloys, the metals considered in this chapter have other applications. Nickel is the most versatile: it is used in copper-nickel alloys for chemical plant, coinage and condenser tubes; in nickel bronzes which are particularly resistant to steam erosion; in nickel silvers for decorative purposes; in the pure form for electroplating and as catalyst in chemical reactions.

Chromium, although less ubiquitous, has a second important use — as an electroplating material to provide a hard surface or merely for decoration. This application consumes a small tonnage, however. A pound of chromium will plate hundreds of car bumpers — an economy that is not reflected in the price of the finished article.

Molybdenum, vanadium and tungsten are largely limited to their use in the steel industry. But the other applications of molybdenum and tungsten, although consuming small tonnages, are important to us — electric lamp filaments, radio tube elements, electrical contact points.

Cobalt is used not only in jet engines, but in similar alloys to give wear resisting surfaces on valves and tools. The metal is also an essential component of certain alloys having special magnetic properties.

It should be evident that the production of alloy steels has involved a considerable amount of metallurgical research. Sometimes a discovery was made more or less accidentally. For example, in 1913, Harry Brearley of Sheffield noticed that a steel he had made and thrown on the scrap pile did not rust. It contained 14% chromium; and his discovery was the birth of stainless steel. On the other hand, the effect of adding as little as 0.001% of boron to steels to produce a hardenable material was only determined after purposeful research. (However, the value of boron is still disputed by some.)

There have, too, been piles of papers published upon the fundamental aspects of alloying. What—exactly—happens when a steel is slowly cooled? What is different in its behavior when it is quenched? How do the internal constituents of steel affect the mechanical properties? What happens when a steel is tempered? Attempts have been made to answer hundreds of questions of this sort. Time will undoubtedly sort out these contributions; but the value of even the already forgotten researches should not be underestimated. For when future generations review the first half of the 20th century and are forced to the conclusion that we have not used our metals wisely (and not because of misguided purchases), they will, however, have to admit that we tried to understand our metals well.

CHAPTER 13

New Metals and Alloys

THE ADVENT of nuclear power reactors has precipitated unique metallurgical problems, for the components of the reactor may require (in addition to adequate mechanical properties) a peculiar corrosion resistance, specific neutron-absorption characteristics and resistance to radiation damage. Further, the fuels for the reactor must not only be fissionable, but (in all but homogeneous reactors) they must be capable of being fabricated into suitable shapes, have adequate resistance to the corrosive matrix that surrounds them, and possess mechanical stability at operating temperatures.

Briefly, a nuclear power reactor comprises the following essential components:

1. The fuel; a fissionable material
2. A moderator that surrounds the fuel. Its purpose is to slow down fast neutrons that are released during fission, so that they may be more readily recaptured and hence keep the fission reaction going.

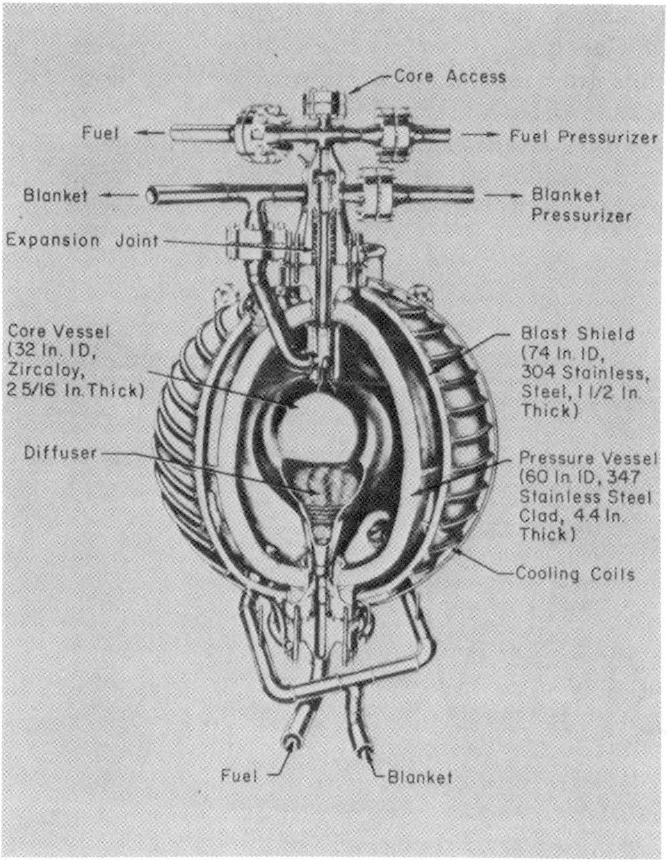

CUTAWAY SECTION OF TYPE HRE-2 REACTOR VESSEL, USING LIQUID FUEL.

Courtesy: U. S. Atomic Energy Commission

3. An agent to transfer the heat of the fission reaction to some form of heat engine; say, a steam-raising plant.

The fuel of an atomic reactor is uranium or thorium. In homogeneous reactors there is no constructional problem of making a fuel element, for the fuel is dispersed in a circulating slurry. However, homogeneous reactor fuels give additional problems of corrosion; and it is not easy to find

MASS-PRODUCED REACTOR (ONE EACH TWO WEEKS) USED FOR TRAINING PURPOSES AND FOR PROCESS CONTROL.
Courtesy: Aerojet-General Nucleonics and Lead Magazine

suitable media to carry the fuel. In heterogeneous reactors, the fuel is in the form of solid elements. Unfortunately, both uranium and thorium are weak at operating temperatures. The strength can be improved by alloying; but, even then, the material will not withstand the highly corrosive attack of the heat-transferring agent or the moderator with which it is in contact. Consequently, the fuel is sheathed. However, the sheathing material must be one that will not absorb too many neutrons; it must have a low thermal neutron absorption cross section, or else fuel will be needlessly used to produce neutrons

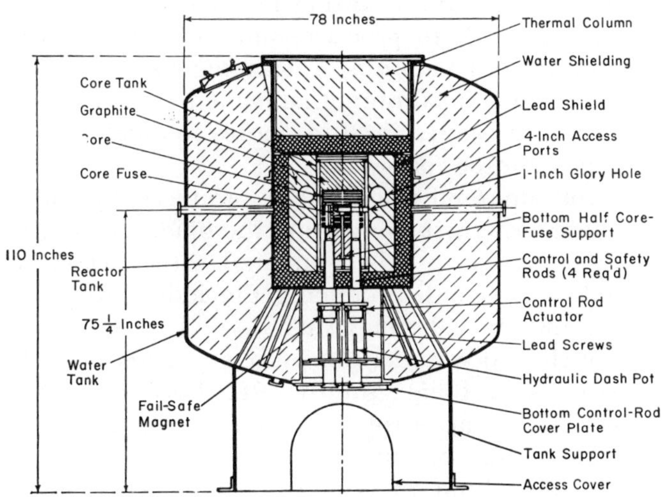

CROSS SECTION OF REACTOR SHOWN ON FACING PAGE.
Courtesy: Aerojet-General Nucleonics and Lead Magazine

that are not contributing to the heat output. Aluminum and zirconium appear to have the most promising properties.

Zirconium of fair purity was first made about 1914. The metal made its commercial debut as an element added to magnesium alloys (about 1% generally improves mechanical properties). However, whereas additions to magnesium alloys could be made of zirconium in the form of compressed powder pellets of not too critical purity, atomic power plants demand a high-purity, ductile material. Now, zirconium is a difficult metal; its affinity for nitrogen, oxygen and hydrogen causes it to absorb these elements from the air and from water vapor in the air; and the resulting metal is brittle. The metal is difficult to produce, since zirconium oxide (the source of zirconium) is extremely stable. Consequently, the metal cannot be made by conventional reduction processes, and a more roundabout approach has to be followed.

Zirconium (and titanium) is produced by one of two methods: first, converting the oxide to chloride and then reducing this compound with magnesium or sodium; second, producing the iodide from the oxide, and then decomposing it.

The chloride-magnesium process was first used by Troost about 1865; and in 1914 Lely and Hamburger made malleable zirconium pellets by the chloride-sodium technique. (Deville, in 1854, used a chloride-sodium process for making aluminum.) The iodide process was introduced by Van Arkel in 1925. During the last 25 years, William Kroll and his associates have developed the chloride-magnesium processes for titanium and zirconium, and have made commercial production a reality. The iodide method produces a purer metal, but it is too expensive for commercial consideration. The chloride method produces a sponge that may be remelted in an arc furnace (using a cooled copper hearth and an inert atmosphere or a vacuum), and the resulting cast ingots are forged into appropriate shapes.

However, the difficulties of producing ductile zirconium on a commercial scale have led to the introduction of a third process which involves powder techniques. Powders of high purity zirconium, or zirconium hydride, are cold pressed, and then sintered to from a solid mass. If metallic zirconium powder is used, sintering may be carried out in a vacuum or under an inert gas, such as argon or helium. But if zirconium hydride powder is used, firing *must* be carried out in a vacuum, in order

WITH 44-ACRE COVERAGE, THIS PLANT AT OAK RIDGE MAKES URANIUM-235 FOR ATOMIC ENERGY PRODUCTION.
Courtesy: U. S. Atomic Energy Commission

that the hydride may decompose to form the metal.

For use in power reactors, zirconium must be free from hafnium, with which it occurs in its natural state. The presence of 2% hafnium reduces the efficiency of zirconium as a low neutron capture material by a factor of approximately 10. The first production of hafnium-free zirconium was at the Oak Ridge National Laboratories in 1949.

There are many possible alloying elements to improve the strength of zirconium without affecting its low neutron capture characteristics. They include

VESSEL CONTAINING TITANIUM SPONGE PRODUCED BY REDUCTION OF TITANIUM CHLORIDE.
Courtesy: Titanium Metals Corporation of America

such common metals as iron, nickel, cobalt and the more unusual elements, (niobium and tantalum, for example).

Uranium as a fuel for heterogeneous reactors is most likely to be made in alloy form. Pure uranium undergoes a continual physical growth and will distort and weaken due to radiation. It is hoped that uranium alloys will overcome, or at any rate alleviate, this difficulty; and consequently, research is in progress whose purpose is to study the physical metallurgy of uranium alloys — that is, to find out how uranium and its alloys behave when they are irradiated. How impossible it might have seemed only 20 years ago that uranium would be available for large-scale uses; and that techniques would be developed for handling radioactive material so that it might, for example, be examined under the microscope and mechanically tested.

No account of nuclear metallurgy would be complete without mentioning the realization of the 7th century alchemists' hope — the transmutation of metals. Transmutation began to be regarded as more than a dream with the discovery of radioactivity, which indicated that transmutation occurs in nature. The possibilities of artificial transmutation were opened up with Rutherford's work of 1919, when he bombarded nitrogen gas with alpha particles and produced entirely new materials — hydrogen and an isotope of oxygen. At that time the alpha particles were obtained from a natural source, a radioactive element. But in 1932, Cockcroft and Williams produced artificial charged particles (protons), which, when allowed to bombard lithium, produced helium. Light elements are more easily decomposed by bombardment than heavy ones; but the advent of the cyclotron, and the development of other more awful accelerators permit the production of charged particles with enormous velocities that are capable of disintegrating the nuclei of heavy elements.

In the type of reaction just considered, transmutation only proceeds as long as bombardment continues. But in 1934, Curie and Joliot bombarded aluminum with alpha particles and found that the reaction continued after bombardment. An isotope of phosphorus was formed, which subsequently broke down to form silicon, and in the process produced another type of emanation — the positron. This is known as "artificial radioactivity."

Now, when uranium is bombarded with neutrons, fission occurs: the uranium nucleus breaks down. But among the fission products are more neutrons; and hence the process is regenerative. This, of course, is the principle behind the reactions for producing atomic power. Another fission product is plutonium which, when first produced in 1941, was regarded as a synthetic element. However, Nature rarely lets us put one over her, and in 1942 the element was detected in pitchblende. The nucleus of plutonium itself may be split to generate a terrific amount of energy; and for this reason the element was used in the atom bomb.

In the previous pages, titanium has been associated with zirconium. This is reasonable, since they are similar metals in many respects, and are produced in much the same way. Of the two metals, zirconium

REMELTED TITANIUM SPONGE IS MADE INTO A 2-TON INGOT.
Courtesy: Titanium Metals Corporation of America

is the more reactive and seems to offer the more practical difficulties. However, titanium is bad enough; but its usage seems to be imperative. At the ridiculous speeds at which aircraft (of impossible sizes) are already flying, the structure must withstand more severe loads; and, of course, parts of the airframe and skin become hot due to their proximity to hot gases, or due to air friction. Now, aluminum alloys are not particularly strong, especially at temperatures above about 300°F.; and, additionally, they creep. Therefore, an alternative material was sought. Titanium alloys, in addition to what appeared to be adequate creep resistance up to temperatures about 800°F., have favorable properties for aircraft construction: they are strong and light; fatigue resistance is very good; corrosion resistance is most satisfactory; and impact properties are superior to those of aluminum.

However, the difficulties of extracting and fabricating titanium are reflected by the 1958 prices of the product at various stages of manufacture:

	Approximate Price per Lb. of Titanium Content
Ilmenite concentrate (TiO_2) ..	$ 0.027
Titanium sponge	3.00
Titanium sheet	10.60

These figures pertain to titanium made from the chloride. Iodide titanium (not fabricated into shapes) runs about $100 per lb.

The necessity of using titanium and its alloys is shown by recent production figures: in 1948, only 2 tons were made; in 1956, about 14,000 tons were

TITANIUM INGOTS, PREPARED FOR SHIPPING.
Courtesy: Titanium Metals Corporation of America

INDIUM, IN A VARIETY OF SPECIAL SHAPES, FOR TRANSISTORS, RECTIFIERS, BEARINGS, LOW MELTING-POINT ALLOYS.
Courtesy: Consolidated Mining and Smelting Company

produced. The price has shown some reduction; but there appears to be no reason to suppose that the price of titanium sheet will drop below about $2.50 per lb. — which is more than three times the present price of stainless steel.

Perhaps the best barometer to indicate the current interest in titanium is the wealth of papers on all aspects of its metallurgy published in contemporary journals, and the immense sums of money that industry and government are spending on its development.

During recent years an increasing interest has been shown in "scarce metals" — metals whose annual production lies between 100 lbs. and 100 tons. Although they are consumed in small quantities, they have peculiar and important uses. Among them are the low melting point metals — gallium, indium and caesium. The low melting points of gallium and indium are used directly in thermometric and fusible plug applications; and for "doping" germanium and silicon for transistors. Caesium's low melting temperature is not utilized; but the metal is used in photoelectric cells as an electron emitter. Tellurium and germanium are semiconductors, although only germanium is used in transistors. Tellurium is used as a minor addition to improve the machinability of copper-base alloys, and is used in the manufacture of some cast irons. Rhenium is used for special electrical purposes (its corrosion product, the oxide, is conductive) and in some types of electronic tube. Of all the scarce metals, rhenium is believed to have the most spectacular future.

CHAPTER 14

The Fabrication of Metals and Alloys

METALS WERE first shaped by beating them. Although the stone implement has been replaced by a more precise hammer and muscle has been superseded by machines, the method has not changed in principle. About 6000 years ago, metals were first cast; and we cast them still. But the techniques of forging and casting have developed over the years; and new methods of fabrication – such as making shapes from powders – have been introduced fairly recently. It is the purpose of this chapter to describe, very briefly, contemporary processes of making metals into marketable forms.

If a metal part can be fabricated by casting it into shape, then it is reasonable to suppose that it may be made comparatively cheaply, because subsequent shaping operations are avoided. It is not, of course, as simple as that, since it might be possible to make a suitable forging of a cheap alloy, while a cast part might demand a more expensive material, or a more elaborate machining schedule. However, in terms of

the number of manufacturing processes involved, casting is the simplest shaping operation. While castings contain imperfections of structure that are, as it were, ironed out by a subsequent forging operation, the number of commodities that *can* be used in the as-cast condition is surprisingly large. One only has to look under the hood of a car to appreciate this: the cylinder block and head are of cast iron (piston rings may be, too); if your car is fairly new, the pistons are of a cast aluminum alloy. Fuel pump, carburetor, horn assembly, grilles are probably of cast material. It is not uncommon for the crankshaft to be made of cast iron.

While many castings can be made in sand (indeed, many *must* be, since alternative methods are often neither appropriate nor economical), more recent techniques of gravity die-casting, pressure die-casting and centrifugal casting have widened the scope of the as-cast component. When a metal is cast in sand, the mold has to be destroyed when the casting is removed; a new mold has to be rammed up around the pattern before another casting can be made. When a large number of castings have to be made from the same pattern, sand casting therefore involves excessive labor costs. The more economic alternative is die-casting, where the mold is made of a non-destructible material — usually iron or steel. In gravity die-casting the metal is poured into the

---

CONTINUOUS CASTING MACHINE, SHOWING (TOP TO BOTTOM) LADLE, MOLD, WATER-SPRAYS, PINCH-ROLLS, CUT-OFF, AND PIT.

Courtesy: Atlas Steels Limited, Welland, Ontario, Canada

iron or steel mold; and when it has solidified the mold is taken apart, the casting removed, and the mold can immediately be used again. At the turn of the century this process was further improved by injecting the metal into the mold under pressure (pressure die-casting), and the method is excellent for the mass production of articles whose dimensions must be precise. As well as a greater dimensional accuracy, die-cast parts (gravity or pressure) have a better surface appearance, since the casting is formed against the machined surface of metal, instead of against particles of sand.

Centrifugal casting is used particularly in the manufacture of hollow shapes of circular cross section. Molten metal is poured into a steel or iron mold which rotates at a speed up to 1000 rpm; the metal is spun to the inner surface of the mold where it solidifies. If for example, the mold were a hollow tube, then a casting would be formed whose outside diameter is the same as the inside diameter of the tube, while the inside diameter of the casting depends upon the amount of metal fed in. Cast iron water pipes are often made in this way. Even when hollow castings are not required, the method may be used, for it has the advantage of producing a very "tight," sound casting.

In recent years the process of "shell-molding" has become popular, particularly for the production of light-metal castings. A metal pattern is first made, and a mixture of sand and resin is sprayed onto it. This shell is stripped and cured, and forms the mold — which is, of course, used only once.

This is, perhaps, the appropriate place to mention

the technique of continuous casting. The purpose of this method is not to produce a finished casting, but to provide material for forming operations. Continuous casting, as the name implies, produces a continuous ingot of material, instead of the conventional "piece" ingots. The technique involves a bottomless ingot mold into which metal is continuously poured; the metal solidifies in the mold, is drawn through, and is quenched as it emerges. A torch cutting unit, geared to the descending ingot, cuts it into suitable lengths.

The method was first applied to copper, aluminum and magnesium, where it led to substantial economies; and it is now in use in commercial steelmaking operations. The advantages of the method are particularly desirable in steelmaking practice. In conventional methods, ingots frequently have to be made of very large section in order to keep their number down to a manageable figure. Consequently, in rolling operations, the amount of reduction (to produce, say, thin strip or bar) is very great; and while this reduction improves the quality of the steel, beyond a certain point it is not economical. Continous casting allows the production of material of small section, and therefore completely eliminates the usual casting pit of ingot molds, as well as reducing the amount of rolling. Secondly, conventional steel ingots usually have a fairly substantial crop taken off their tops (where the metal is porous or contaminated) ; sometimes, a crop is taken off the bottom of an ingot, too. Clearly, when the ingot is in one long piece, cropping is avoided.

The mechanical properties of metals are improved by "working": toughness and ductility are particularly enhanced. So, in addition to producing shapes that might be difficult to make as castings (wire, for example), a working operation results in a sounder product. Most metals harden when they are worked at normal temperatures; and while this effect may be used to advantage (in increasing the strength of certain types of stainless steel, for instance), the greater strength demands that more power is supplied to the deforming process. Further, cold working may eventually put the metal in such a brittle condition that it fractures during fabrication. Consequently, working processes frequently involve either annealing treatments between successive forming operations, or shaping the metal while it is hot. Cold working is generally restricted to operations where the metal is ductile, and the amount of deformation can be accommodated; to the production of a shape that cannot be strengthened by subsequent heat treatment; or to the manufacture of a product that must have a clean finish. Perhaps the most familiar commodity that is cold worked is the common nail, whose head is cold formed.

Structural shapes (beams, angles, sheet and so on) are generally made by rolling a hot billet of metal through appropriately shaped rolls. The expression "appropriately shaped" is carefully chosen, for it glosses over a multitude of factors that must be considered in setting up a rolling mill. The billet is reduced in size and is shaped by passing it through successive rolls; and while labor and capital costs are lower if the number of stages is minimal, the pos-

SEAMLESS PIPE PRODUCTION. THE BILLET IS PIERCED, SPUN BETWEEN ROLLS AND PULLED OVER A PLUG.
Courtesy: American Iron and Steel Institute

sibility of the billet tearing during rolling is increased if the amount of reduction at each stage is too great.

Wire production uses as its starting material rod from the rolling mill. The rod is cleaned, pointed at one end, and drawn (hot or cold) through a series of dies. The dies must be made of a hard material that will keep its shape: certain kinds of cast iron, chromium-plated steel, tungsten carbide are commonly used.

Many nonferrous metals may be extruded into shapes. In principle, the process is as simple as squirting toothpaste from a tube; and, appropriately

EQUIPMENT FOR RADIOGRAPHIC INSPECTION.
Courtesy: Steelways

enough, the toothpaste tube itself is made by an extrusion process. A hot metal billet is placed in a container which has, at one end, a die whose shape determines the cross-section of the extruded product. At the other end, a hydraulically operated plunger is applied, and metal is forced through the die. Alternatively, the container may be closed at one end, while the plunger itself contains the die.

While this type of extrusion process uses hot metals, some cold metals can be formed by impact extrusion. A slug of cold metal is placed in a die, and is hit by a punch. Metal is forced into the gap between the die and the punch. The process is limited to soft metals like lead, tin and aluminum.

Although most nonferrous metals may be made into tubular products by extrusion processes, steel is generally made into tube by other techniques. The importance of steel tube in the world today is obvious enough — in oil and gas lines, domestic

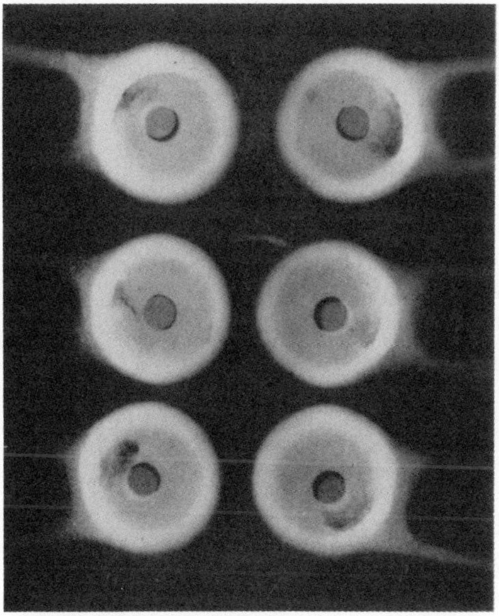

RADIOGRAPHS OF STEEL CASTINGS. DARK REGIONS IN LIGHTER AREAS INDICATE CRACKS, VOIDS, ETC.

Courtesy: Steelways

services, construction, furniture. Seamless tube is made by piercing a hole extending part way through a hot billet, inserting a mandrel, and pushing the billet (by applying a force to the mandrel) through dies. Or, the hot billet, with the inserted mandrel, is spun between inclined rolls that draw the metal between them. Such tube does not have the longitudinal seam that characterizes other manufacturing processes.

Other tube and pipe is made by forming metal strip between rolls that gradually bend up the edges until a circular section is produced, and the edges touch. They are then welded, usually electrically. The welding machine comprises a circular wheel made of two copper disks which are insulated from each other. The disks run on either side of the longitudinal joint in the formed metal; a heavy current is passed into one disk, through the metal via the seam, and out through the other disk. The heat developed is sufficient to weld the seam.

The processes described so far are used, substantially, to fabricate metal in lengths – angle, pipe, wire, and so on. However, many forming operations are concerned with shaping a piece of metal into something with a personality of its own, not a commodity that is produced by the mile. Forging processes of many kinds are used to make a range of products from dimes to drive shafts. Forging involves hammering or squeezing hot or cold metal to produce a required shape. Drop forging is done with hammers; press forging uses the squeeze. The metal to be fabricated may be partly or wholly enclosed by forming dies that give it the desired shape.

Towards the borderline of the area covered by the term "forging" lie the operations known as cold pressing and deep drawing. These manufacturing techniques involve shaping a sheet of metal by pressing it between dies. Cartridges, lamp sockets, auto bodies, steel sink units are made by these sorts of process.

Some metal sheet lends itself to other forming processes. For instance, aluminum and aluminum alloy sheet may be stretch pressed for aircraft skins. The metal, pulled tight over a form, is pressed, and stretched plastically to attain the shape of the form. Aluminum sheet may also be "spun": a disk of metal is rotated and forced either into or around a shaped chuck by a blunt tool. Saucepans are made in this way.

There are metals that defy the conventional forming techniques because their melting temperatures are high. Consequently, an ingot cannot be cast; and that in turn might be expected to eliminate the types of working process that have been described. Such metals, in powder form, may be compacted and sintered. Possibly the earliest application of this sort of process was in the production of tungsten wire for lamp filaments. Tungsten powder is produced by the normal method of extracting metallic tungsten from its ore. This is compacted into short rods and sintered. The sintered material is treated by a process of "swaging" which puts a point on the rod; and then it may be drawn through dies to form wire. Carbides of tungsten, molybdenum, tantalum and other metals are made into shapes by powder metallurgical methods in the production of machine tools and dies. Although castings can be made with difficulty, they are usually too brittle to be of use. Therefore, carbides are mixed with powder of a tougher metal (usually nickel or cobalt), and the tool is fabricated by pressing and sintering.

However, powder metallurgical techniques are not restricted to metals and compounds of high melting point. Porous bearings into which oil is injected under pressure and which therefore need no further lubrication are made by powder methods; filters may be produced in a similar way. Products that consist of a mixture of metal and non-metal (for instance, copper and graphite — again, a bearing material) are made by the sintering of powders.

Certainly since the time of the Egyptians has it been appreciated that it is often simplest to make some articles by joining several parts together. Many statues were cast in parts which were fastened together by pins and joints. Later, rivets and bolts became more common. And, in recent years, welding processes, although not new to our times, have developed tremendously.

Metals may be welded without actual fusion. The method used by the blacksmith involves heating the

LABORATORY FURNACE FOR MELTING IN VACUUM OR INERT ATMOSPHERE.

Courtesy: National Carbon Company

parts to be joined, and hammering them together. The metal does not melt; but the metal crystals at the interface grow into each other, and so effect a joint. The method still has important uses; and because the metal is not melted, there is no cast structure at the joint. Consequently, the disadvantages of castings, which are present to some extent in all fusion welds, are avoided. Pressure welding is used, for example, in the production of "Alclad" sheet. This is an aluminum alloy that is sheathed by pure aluminum: a billet of alloy is sandwiched between sheets of pure aluminum and the whole thing is rolled. The result is a sheet that has the strength of the alloy together with the greater corrosion resistance of pure aluminum; and there is no doubt that the aluminum and the alloy are firmly bonded by this pressure welding process.

However, with the disappearance of the blacksmith's forge and the advent of complicated engineering design, fusion welding has become more familiar. In this process the parts to be welded are held together and their junction is melted — with, or without the addition of more metal. The weld has the structure and properties of a cast metal. The source of heat may be a gas flame, or an electric arc, or the passage of an electric current through the work. The "Thermit" process (introduced at the beginning of the century) was seen quite frequently, too, when tram lines were more common. In this method a sand mold is made around the parts to be joined, and a mixture of aluminum powder and iron oxide contained in a crucible is ignited immediately above the joint. The hot metal is then

A COCKCROFT-WALTON ACCELERATOR — TO SPEED UP ATOMIC PARTICLES FOR ATOM-SMASHING EXPERIMENTS.

allowed to flow into the sand mold; it fuses the metal parts; and the metallic iron produced in the reaction fills the gap, while the slag produced floats to the surface.

In welding processes fluxes are often used to react with or dissolve any oxides that are formed. A slag is formed which (it is hoped) floats up out of the weld metal. Some metals, such as aluminum and titanium, are extremely reactive and defy fluxing methods; the bare surfaces to be welded cannot be kept clean. Consequently, electric arc welding of these metals is performed in an inert atmosphere by maintaining an envelope of argon or helium gas around the arc.

Soldering and brazing processes may, perhaps, be described as being intermediate between pressure welding and fusion welding. In both of these processes the parent metal is not fused, but the material used for making the actual joint is molten. It is generally agreed that the mechanism of brazing and soldering involves the formation of an alloy between the solder and the parent metal at the interface. Since the alloy extends into both parent metal and the solder, a joint is effected. Usually the solder and the brazing rods are weaker materials than the parent metal. Therefore, the handyman's advice to keep the amount of solder to a minimum is well founded.

The fabrication of components for engineering purposes often demands that the mechanical properties of the material are adjusted by suitable heat treatment. Machinability of a steel might be improved by annealing; strength may be increased by appropriate quenching and tempering treatments. The subject is such a vast one that it cannot be covered in a book of this kind; and to treat it briefly would only be misleading. The most I dare do is to mention its necessity, so that the reader may give me credit for having remembered about it.

When metals are put into services they are almost always subjected to some corrosive influences. Often, corrosion is the eventual cause of the component's failure; or (if one is luckier) it is the reason for the article's untimely renewal. While careful metal selection can eliminate many corrosion problems, the choice often rests upon the suitability of mechanical properties. Under these circumstances, the metal must in some way be protected from its corrosive environment. The most common method is to give it a coat of paint.

While painting may not be the metallurgist's responsibility, other types of coating do come within his field. A metal may, for instance, be clad with another more resistant metal (as in the case of "Alclad," which has already been mentioned). Or coatings of zinc, chromium or aluminum may be applied by heating the metal in a suitable powder, when diffusion of the powder into the metal surface occurs. Zinc is more commonly applied to iron and steel by galvanizing: the metal is dipped into a bath of molten zinc. Tinplate is made by dipping steel into molten tin. Chromium, nickel and other pro-

A SINGLE CRYSTAL OF SILICON GROWN FOR EXPERIMENTAL PURPOSES IN THIS INDUCTION-HEATED APPARATUS.

Courtesy: Bell Telephone Laboratories

tective coatings may be applied by electroplating. And where all of these methods are not applicable (because, perhaps, the article to be coated is part of an immovable structure), metal may be sprayed on the surfaces to be protected.

Nonmetallic coatings are frequently produced at the factory because they form a good surface on which paint can bite. Methods for producing nonmetallic coatings often involve dipping the metal into a corrosive bath, which reacts with the surface of the metal, leaving behind a tight layer of reaction product, to which the paint can bond. Anodizing (a process that is responsible for the remarkable range of dye colors that may be applied to aluminum) is an electrolytic process whose purpose is to form a thin, tight film of aluminum oxide on the metal surface.

Perhaps this survey has given some indication of the complexity of metallurgical operations that are performed upon many metal articles. And these contemporary techniques of casting, working, machining, heat treating and surface finishing all have their origins in the distant past — casting in a stone mold, working by beating, machining by crude chipping and cutting, heat treatment (such as it was) by simple softening operations or (in iron articles) by quenching and tempering, surface finishing by laborious polishing. There is, as they say, nothing new in the world. But there is, of course, something entirely new about the speed, efficiency and, consequently, the economy of effort and material with which metal goods are made.

The argument that we would be happier if this

facility for making metal articles (and, surely, all articles, since everything we use has seen metal) had not come upon us is a forceful one. The man that preaches the doctrine of the simple life always has a large and attentive audience. But it is, perhaps, worth reflecting, that under the circumstances of the simple life he preaches (when a man killed his food with stone weapons, tilled the soil with stone plows, prepared and cooked his meals in stone vessels), we would rarely be in the mood to listen to his sophistries.

CHAPTER 15

Metallurgical Science in the 20th Century

THIS IS an ambitious chapter heading, which would be more fitting to a ten-volume work — but perhaps we overemphasize contemporary developments, and a mere chapter on the subject may serve the purpose of cutting us down to size. If it serves the additional purpose of indicating some of the present-day problems with which metallurgists are faced, and of surveying a few of the tools with which the problems may be analyzed, then the chapter can justifiably claim to be more than a fag-end.

It should be apparent that by the time large-scale steelmaking had begun, metallurgists were starting to understand some of the whys and wherefores of the techniques they used. If one had to state categorically whose work this gradual enlightenment depended upon, perhaps the best claimant would be Robert Boyle, the 17th century chemist. Boyle, making a distinction between elements, compounds and mixtures, laid the foundations of modern chemistry, clearing away a decaying alchemical philosophy and an inflexible and useless phlogistic theory that had outlived its usefulness.

The hypotheses of the more advanced scientists of the middle and late 17th century are, at any rate, sufficiently akin to 20th century ideas that we can read the work of these scientists and understand what their intentions were. For example, Réaumur, who worked during this period published his "Memoirs on Steel and Iron" in 1722. It includes not only practical details, but explanatory theories of such processes as malleabilizing cast iron and heat treating steel. The work (translated by Sisco, and published under the auspices of the American Iron and Steel Institute, 1956) shows a startling insight into ferrous metallurgy, and an ability to explain effects in terms of what were then modern scientific principles.

While the skeptic might suggest that the art of getting metals from their ores could have jogged along quite well without the intrusion of the scientist, it is unlikely that he would maintain that contemporary processes of fabrication and treatment could have been born out of anything but scientific discovery. Much of the metallurgical research conducted over the past fifty years has been concerned with finding out how metals behave, and in trying, with rather less success, perhaps, to find out just what a "metal" really is.

In the same way as we know a politician by his color, his party characteristics and his friends, so a metal may be identified by its luster, its properties, and the other metals to which it shows some similarity. A hundred years ago, the metallic state was understood only in terms of strength and ductility, properties of conducting heat and electricity, lustrous appearance and chemical activity. The crystalline structure of a metal was suspected on account of its granular appearance when it was broken with a brittle fracture, and also because of the large crystals which are sometimes seen at the surface of a solidifying metal.

The then recently introduced technique of examining rock sections by transmitted light under the microscope, which was successfully applied by Nicol, seemed to have no application to metals. Generally, metal sections, however thin, were found to be quite opaque; and although Faraday made some interesting experiments with one of the exceptional metals, gold, by viewing a thin leaf by transmitted light, his observations were of a limited value. Widmanstätten, however, showed rather crudely that a metal could be examined by *reflecting* light from its surface into a microscope. Because of the irregularity of the metal surface no conclusive results could be drawn, and the discovery did not receive much attention. But by using an elaborate polishing technique, involving successively finer abrasives, Sorby, in 1863, was able to produce a mirror-like finish. Under the microscope, though, the structure of most metals prepared in this way is concealed by what is still commonly agreed to be an "amorphous layer." During polishing, it may be supposed that the immediate surface of the metal becomes so hot that it melts, and then cools very rapidly to produce an amorphous, or at any rate, a very minutely crystalline, film. This is only a few atoms thick; it is quite structureless under the microscope; and it may be removed by dissolving it in a mildly corrosive solution. The underlying surface, characteristic of the crystallinity of the metal, is then revealed.

Sorby's technique of microscopical examination provides one of the most powerful tools of the metallurgist. The crystals of a metal are outlined by their boundaries — for these are more active than the crystal proper, and are thus heavily attacked by the corrosive etching solution; the constituent parts of an alloy are differently affected; while under some conditions, characteristic colors may be produced. Given a little information about the composition of an alloy (and it is sometimes possible, though rarely permissible, to guess at this) a competent metallographer may be able to interpret from what he sees under the microscope the history of the material: the way in which it has been fabricated; the nature of its heat-treatment; the possible causes of failure.

Polishing and etching methods have advanced since Sorby's day, making the technique so valuable that even in these days of electron microscopes, X-rays, radioactive tracers, and all the rest of the contemporary metallurgist's fussy equipment, microscopical metallography is not only standard practice in industrial laboratories, but contributes generously to metallurgical research.

The metallurgical tangle was unravelled a little more — and, indeed, is still being unravelled — by studying the changes that can occur in the physical properties of a metal or alloy as its temperature is altered. These may be changes in internal energy, manifested by an evolution or absorption of heat;

changes in volume, or in magnetic or electrical characteristics. By tying these observations to microscopical appearances, it has been possible over the last fifty years to deduce with some accuracy how metallic materials behave under different imposed conditions of temperature; and this, in turn, has suggested new treatments and uses. Constitution diagrams have been evolved by this sort of experimentation, which indicate the make-up of alloys of different compositions at different temperatures.

A great impetus was given to metallurgical science when, in 1912, Laue suggested an analysis of crystal structure by means of X-rays — which had been discovered by Röntgen some seventeen years previously. Experimental work by Friedrich, Kipping, the Braggs and subsequently many others, established a completely new approach to crystal structure investigation, and the techniques were very successfully applied to metals.

While X-rays may be used to detect flaws in metals in the same way as a doctor detects flaws in our bodies, the analysis of crystal structure by X-rays depends upon an entirely different principle. Roughly, it is this: when a beam of X-rays falls upon a regular array of atoms (as are present in a crystal) the X-ray beam is diffracted in several directions. The diffracted beams can be picked up on a photographic emulsion or by Geiger counter, and the angle through which they have been diffracted, with respect to the incident beam, is easily calculated. These angles depend upon the pattern to which the atoms of the crystal conform: hence the atomic pattern may be deduced. So, the use of

X-ray diffraction methods allows the metallurgists to "see" — but indirectly — the atomic arrangements in metals, and how they are affected by alloying, heat-treatment and fabrication.

But the metallurgist is not restricted to beams of light (in the microscope) or to beams of X-rays (in diffraction studies). He can, if he has the fortitude, use electron beams. The application of the electron beam is twofold: first, to use it rather like a light beam, and determine topographical features; second, to use it like an X-ray beam, and interpret atomic patterns. But in both of these applications the small wavelength of electrons compared to light or X-rays, permits a much greater resolution. Magnifications using an ordinary light microscope are limited to about two thousand times: electron microscopy gives magnifications ten times higher. In X-ray diffraction work, very small aggregates of atoms pass undetected: electron diffraction techniques are more likely to pick them up.

These experimental tools — and, of course, many others both clumsy and subtle — have given the metallurgist the chance to try to determine the reasons behind the behavior of metals and alloys. And, over the past twenty or thirty years, the metallurgist has been joined by the physicist in his investigations into metal structure. Perversely enough, though, it is not the massive things seen under the microscope, nor the regularity of atomic arrangement revealed by X-rays, that is presently occupying one important center of interest. For although a crystal of metal appears homogenous when viewed through the microscope, and X-ray diffraction studies indicate that the crystal is made up of a regular three-dimensional array of atoms, neither microscope nor X-rays can reveal that the atomic pattern may be locally upset by gaps and other irregularities. Yet, the contemporary view is that this, in fact, is so. Perhaps the most convincing argument in favor of theories that propose imperfections in what appear to be perfectly regular crystals is this: when a crystal of metal is broken, one imagines that atoms of metal must be pulled away from neighboring atoms. But calculations show that in order to pull atomic neighbors apart, a force must be applied to the crystal that is hundreds of times greater than is actually required in practice. If, however, one presumes that "dislocations" are present in the atomic pattern, whereby the regularity of the atomic arrangement is locally destroyed, then the calculation agrees with experiment. Of course, at this point the skeptic quite reasonably objects, and asks upon what grounds can the "local imperfections" be justified. Although there are many phenomena that convince metallurgist and physicist, a more normal soul is probably won over by the following argument and experiment: if the dislocations are only local, then there should be a fair quantity of "undislocated" metal between the dislocations. If a piece of metal could be obtained so thin that it was smaller in section than the average distance between dislocation, then it should prove to be exceptionally strong. In breaking such a piece of metal, neighboring atoms must be torn apart, with no local disturbances to send them on their way. This is, in fact, the case. In recent years, metal "whiskers" have been grown which are sensibly free from dislocations, and these have the strength and other mechanical properties that are approximately those calculated for a perfect atomic array.

While dislocation theories are having a pronounced effect on the metallurgists' interpretations of such phenomena as the strength of metals, the way atoms move through crystals — phenomena that we may think of as mechanical in their nature — many other aspects of metal structure fall outside the bounds of this sort of theory. For while in considering the movement of atoms, the atoms may be assumed to resemble little balls, neatly arranged and happily spherical, it often becomes necessary to consider the constitution of the atom itself. For instance in the matter of the electrical conductance of metals, the behavior of the electrons associated with the atoms is critical. In fact, it is the way in which the electrons behave that seems to principally distinguish a metal from a non-metal.

In view of such a sharp distinction, it is not surprising that a consolidated "electron theory of metals" has been built — built upon the foundations of quantum mechanics, which were themselves laid at the turn of the century. This is not to suggest that the theory offers a complete picture in which the behavior of metal electrons can always be clearly seen: for instance, the effect known as superconductivity has yet to be convincingly explained. When certain metals are cooled to very low temperatures their electrical resistance becomes zero — which means, of course, that an electric current in a closed circuit of a superconducting metal keeps on circulating and will never die at these "superconducting temperatures."

Again, the explanations offered to account for the magnetism of iron, nickel, cobalt and gadolinium (as opposed to the non-magnetism of metals like copper and lead) are not yet altogether acceptable. Yet, in their defense it must be pointed out that the existing theories have led to predictions that have proven to be true. Only recently it was announced that an alloy of manganese and bismuth — both metals are nonmagnetic — can be made into a magnet which has exceptionally high permanence; and one supposes that this discovery was not a shot in the dark.

Contemporary metallurgical research is not easily classified. It ranges from the naively empirical to the inscrutibly theoretical. While some metallurgists are engaged in dunking an alloy into a corrosive solution simply to see how long the alloy will last, the boys in the next room (who probably cannot distinguish a chromium-plated teaspoon from a silver one) are scratching out their latest ideas on semiconductors. But it is a narrow-minded man that scoffs at the first group for their empiricism, or at the second group for their egg-headedness. Both kinds of research, and the vast range that separates them, have their place. Empirical, "let's-try-it-and-see," experiments are remarkable money-savers, even in these enlightened times. And if anyone doubts the repercussions of theoretical research, let him relate a still young atomic theory to the overwhelming effects it has spawned.

Epilogue

The bellows are burnt; the lead is consumed of the fire; the founder melteth in vain.

— JEREMIAH: vi, 29.

IN THE GOOD OLD DAYS of 6000 B.C. a man could be happy with an acre of land, a few flints and a wife, or whatever it was that social conditions allowed. In 3000 B.C. he found it desirable to add bronze tools to the equipment that was to see him through a satisfied life. By 1000 B.C. iron was becoming a war-waging necessity. And by the 19th century A.D. life had become even more complicated — a man's prosperity no longer depended upon his own personal skill, but more largely upon the condition of the country he lived in, upon its agriculture and its industry.

Industry, until 100 years ago, could thrive in a country that had sufficient resources of iron, copper, zinc, lead, tin and coal to feed its factories. A little later, as we have seen, manganese became important to the steelmaking processes; and, as the years went by, the list of commercially important metals grew.

Alloy steels have become necessary for the peculiar conditions that metals must withstand, and in their manufacture many metallic elements must be used. Nonferrous metallurgy has expanded too: aluminum and magnesium and their alloys are in everyday use; the varieties of brass and bronze have become so numerous as to include almost every possible alloying element; zirconium, titanium, uranium are no longer strange names. Today, therefore, there are few metals that could easily be dispensed with.

But the situation does not stop changing. Metals are being consumed that cannot be recovered. The alloying elements in steel become dispersed when mixed batches of scrap are melted down, so that the product contains a harmless (but fairly useless) assortment of elements that cannot conceivably be recovered. The steel put into service rusts away at the rate of thousands of tons each week, dispersing itself (by a nice irony) into the earth from which it originally came. But it disperses itself in concentrations that are too lean to be worked. Zinc from galvanizing, lead from roofings, tin from cans, copper from cables — the list is an easy one to compile, for every metal is slowly going back home. And today, we have, in addition, the more positive reaction in which metals are actually destroyed by atomic fission.

It would be ridiculous to suggest that the position is at all critical. If it were, it would not be nearly so interesting. But it is, perhaps, a pretty conjecture (prettier because it cannot affect any of us) that, in an age to come, metals that have not been fissioned will be back, once more, in the earth in such minute concentrations that they cannot be extracted. Man once more lives an agricultural life. His remaining implements gone, he cooks in earthenware pots, plows with flints, kills with stones. The process has gone full cycle; and if there be a metallurgist, he will fill his time reflecting upon the wisdom of the Prophets.

Index